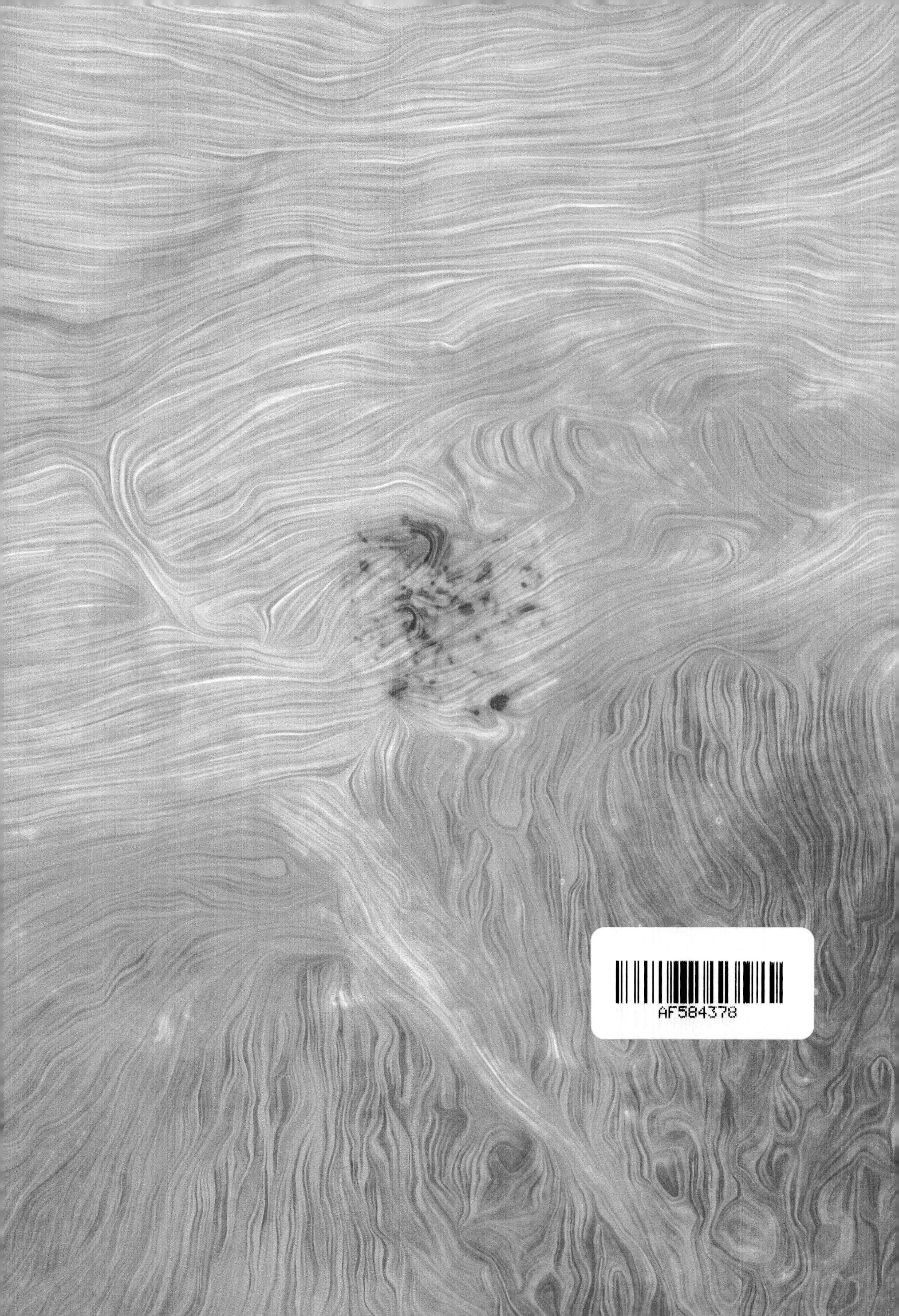
AF584378

Pearson Australia
(a division of Pearson Australia Group Pty Ltd)
707 Collins Street, Melbourne, Victoria 3008
PO Box 23360, Melbourne, Victoria 8012

www.pearson.com.au

First published 2018 by Pearson Australia
2021 2020 2019 2018
10 9 8 7 6 5 4 3 2 1

Lead Publisher: Misal Belvedere
Project Manager: Michelle Thomas
Production Editors: Elizabeth Gosman and Laura Pietrobon
Lead Development Editor: Fiona Cooke
Content Developer: Bryonie Scott
Development Editor: Naomi Campanale
Editors: David Meagher, Ross Blackwood
Designer: Anne Donald
Rights & Permissions Editors: Samantha Russell-Tulip and Jenny Jones
Senior Publishing Services Analyst: Rob Curulli
Proofreader: Diane Fowler
Illustrator/s: DiacriTech
Printed and bound in Australia by Pegasus Media & Logistics

ISBN 978 1 4886 1935 9

Pearson Australia Group Pty Ltd ABN 40 004 245 943

Acknowledgements

The following abbreviations are used in this list: t = top, b = bottom, l = left, r = right, c = centre.

Cover image: Planck Collaboration/ESA/Science Photo Library

123RF: Dario Lo Presti, pp. 1–2; sykwong, p. 109.

Alamy Stock Photo: Melvyn Longhurst, p. 36.

American Association of Physics Teachers: Reproduced from 'How I misunderstood Newton's Third Law', by Hughes M.J. The Physics Teacher 40/381 (2002), http://dx.doi.org/10.1119/1.1511603, with the permission of the American Association of Physics Teachers., p. 51.

Malcolm Cross: p. 91.

Fotolia: Olesia Bilkei, p. 99; petejau, pp. 145–6.

Retrospect Photography: Dale Mann, p. 148.

Science Photo Library: King's College London, p. 188; Claus Lunau, p. 113; Planck Collaboration/ESA, p. i; Trevor Clifford Photography, p. 152; TRL LTD., p. 79; Detlev Van Ravenswaay, p. 136.

Shutterstock: BarryTuck, p. 47; EpicStockMedia, pp. 86–8; kostin77, p. 58; ostill, p. 51; serpetko, pp. 41–2.

Some of the images used in *Pearson Physics 11 New South Wales Skills and Assessment Book* might have associations with deceased Indigenous Australians. Please be aware that these images might cause sadness or distress in Aboriginal or Torres Strait Islander communities.

Practical activities
All practical activities, including the illustrations, are provided as a guide only and the accuracy of such information cannot be guaranteed. Teachers must assess the appropriateness of an activity and take into account the experience of their students and the facilities available. Additionally, all practical activities should be trialled before they are attempted with students and a risk assessment must be completed. All care should be taken whenever any practical activity is conducted: appropriate protective clothing should be worn, the correct equipment used, and the appropriate preparation and clean-up procedures followed. Although all practical activities have been written with safety in mind, Pearson Australia and the authors do not accept any responsibility for the information contained in or relating to the practical activities, and are not liable for any loss and/or injury arising from or sustained as a result of conducting any of the practical activities described in this book.

Contents

Contents

Module 3: Waves and thermodynamics

Module 4: Electricity and magnetism

How to use this book

The *Pearson Physics 11 New South Wales Skills and Assessment Book* takes an intuitive, self-paced approach to science education that ensures every student has opportunities to practise, apply and extend their learning through a range of supportive and challenging activities. While offering opportunities for reinforcement of key concepts, knowledge and skills, these activities enable flexibility in the approach to teaching and learning.

Explicit scaffolding makes learning objectives clear, and there are regular opportunities for student reflection and self-evaluation at the end of individual activities throughout the book. Students are also guided in self-reflection at the end of each module. There are rich opportunities to take the content further with the explicit coverage of Working scientifically skills and key knowledge in the depth studies.

This resource has been written to the new New South Wales Physics Stage 6 Syllabus and addresses the first four modules of the syllabus. Each module consists of five main sections:

- key knowledge
- worksheets
- practical activities
- depth study
- module review questions.

Explore how to use this book below.

Physics toolkit

The Physics toolkit supports development of the skills and techniques needed to undertake practical investigations secondary-sourced investigations and depth studies, and covers examination techniques and study skills. It also includes checklists, models, exemplars and scaffolded steps. The toolkit can serve as a reference tool, to be consulted as needed.

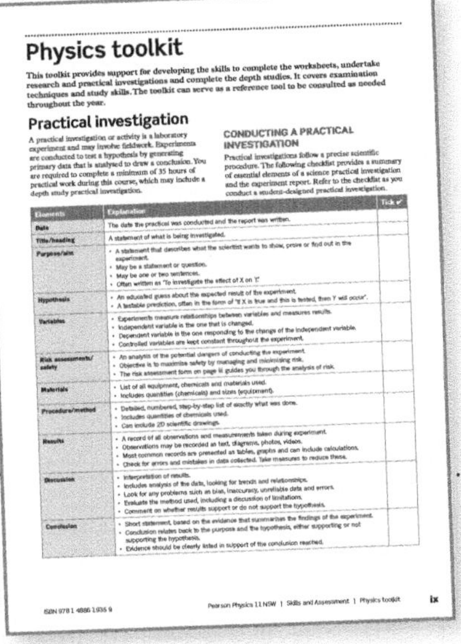

Physics toolkit

This toolkit provides support for developing the skills to complete the worksheets, undertake research and practical investigations and complete the depth studies. It covers examination techniques and study skills. The toolkit can serve as a reference tool to be consulted as needed throughout the year.

Practical investigation

A practical investigation or activity is a laboratory experiment and may involve fieldwork. Experiments are conducted to test a hypothesis by generating primary data that is analysed to draw a conclusion. You are required to complete a minimum of 35 hours of practical work during this course, which may include a depth study practical investigation.

CONDUCTING A PRACTICAL INVESTIGATION

Practical investigations follow a precise scientific procedure. The following checklist provides a summary of essential elements of a science practical investigation and the experiment report. Refer to the checklist as you conduct a student-designed practical investigation.

Elements	Explanation	Tick ✓
Date	The date the practical was conducted and the report was written.	
Title/heading	A statement of what is being investigated.	
Purpose/aim	• A statement that describes what the scientist wants to show, prove or find out in the experiment. • May be a statement or question. • May be one or two sentences. • Often written as 'To investigate the effect of X on Y.'	
Hypothesis	• An educated guess about the expected result of the experiment. • A testable prediction, often in the form of 'If X is true and this is tested, then Y will occur'.	
Variables	• Experiments measure relationships between variables and measures results. • Independent variable is the one that is changed. • Dependent variable is the one responding to the change of the independent variable. • Controlled variables are kept constant throughout the experiment.	
Risk assessments/safety	• An analysis of the potential dangers of conducting the experiment. • Objective is to maximise safety by managing and minimising risk. • The risk assessment form on page iii guides you through the analysis of risk.	
Materials	• List of all equipment, chemicals and materials used. • Includes quantities (chemicals) and sizes (equipment).	
Procedure/method	• Detailed, numbered, step-by-step list of exactly what was done. • Includes quantities of chemicals used. • Can include 2D scientific drawings.	
Results	• A record of all observations and measurements taken during experiment. • Observations may be recorded as text, diagrams, photos, videos. • Most common records are presented as tables, graphs and can include calculations. • Check for errors and mistakes in data collected. Take measures to reduce these.	
Discussion	• Interpretation of results. • Includes analysis of the data, looking for trends and relationships. • Look for any problems such as bias, inaccuracy, unreliable data and errors. • Evaluate the method used, including a discussion of limitations. • Comment on whether results support or do not support the hypothesis.	
Conclusion	• Short statement, based on the evidence that summarises the findings of the experiment. • Conclusion relates back to the purpose and the hypothesis, either supporting or not supporting the hypothesis. • Evidence should be clearly listed in support of the conclusion reached.	

ISBN 978 1 4886 1935 9 Pearson Physics 11 NSW | Skills and Assessment | Physics toolkit ix

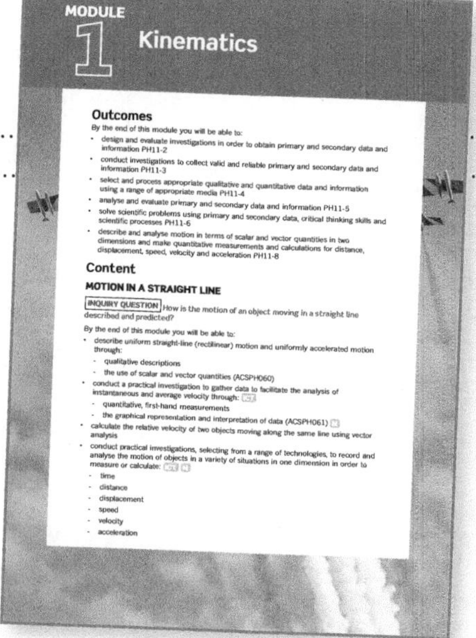

MODULE 1 Kinematics

Outcomes

By the end of this module you will be able to:

- design and evaluate investigations in order to obtain primary and secondary data and information PH11-2
- conduct investigations to collect valid and reliable primary and secondary data and information PH11-3
- select and process appropriate qualitative and quantitative data and information using a range of appropriate media PH11-4
- analyse and evaluate primary and secondary data and information PH11-5
- solve scientific problems using primary and secondary data, critical thinking skills and scientific processes PH11-6
- describe and analyse motion in terms of scalar and vector quantities in two dimensions and make quantitative measurements and calculations for distance, displacement, speed, velocity and acceleration PH11-8

Content

MOTION IN A STRAIGHT LINE

INQUIRY QUESTION *How is the motion of an object moving in a straight line described and predicted?*

By the end of this module you will be able to:

- describe uniform straight-line (rectilinear) motion and uniformly accelerated motion through:
  - qualitative descriptions
  - the use of scalar and vector quantities (ACSPH060)
- conduct a practical investigation to gather data to facilitate the analysis of instantaneous and average velocity through:
  - quantitative, first-hand measurements
  - the graphical representation and interpretation of data (ACSPH061)
- calculate the relative velocity of two objects moving along the same line using vector analysis
- conduct practical investigations, selecting from a range of technologies, to record and analyse the motion of objects in a variety of situations in one dimension in order to measure or calculate:
  - time
  - distance
  - displacement
  - speed
  - velocity
  - acceleration

Module opener

Each book is divided to follow the four modules of the syllabus, with the module opener linking the module content to the syllabus.

Key knowledge

Each module begins with a key knowledge section. The key knowledge consists of a set of succinct summary notes that cover the key knowledge set out in each module of the syllabus. This section is highly illustrative and written in a straightforward style to assist students of all reading abilities. Key terms are bolded for ease of navigation. It also serves as a ready reference for completing the worksheets and practical activities.

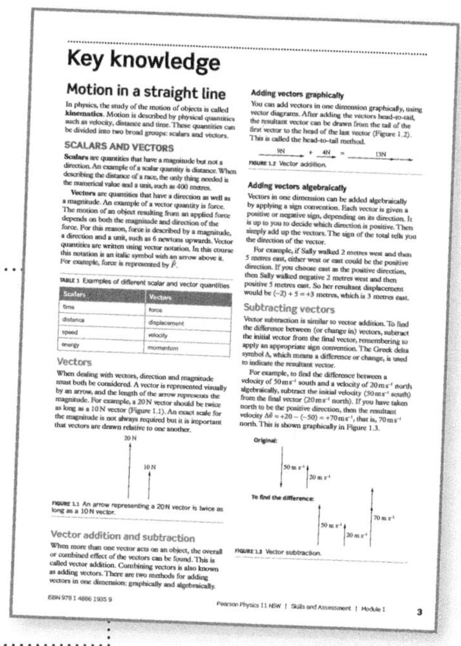

Key knowledge

Motion in a straight line

In physics, the study of the motion of objects is called **kinematics**. Motion is described by physical quantities such as velocity, distance and time. These quantities can be divided into two broad groups: scalars and vectors.

SCALARS AND VECTORS

Scalars are quantities that have a magnitude but not a direction. An example of a scalar quantity is distance. When describing the distance of a race, the only thing needed is the numerical value and a unit, such as 400 metres.

Vectors are quantities that have a direction as well as a magnitude. An example of a vector quantity is force. The motion of an object resulting from an applied force depends on both the magnitude and direction of the force. For this reason, force is described by a magnitude, a direction and a unit, such as 6 newtons upwards. Vector quantities are written using vector notation. In this course this notation is an italic symbol with an arrow above it. For example, force is represented by $\vec{F}$.

TABLE 1 Examples of different scalar and vector quantities

Scalars	Vectors
time	force
distance	displacement
speed	velocity
energy	momentum

Vectors

When dealing with vectors, direction and magnitude must both be considered. A vector is represented visually by an arrow, and the length of the arrow represents the magnitude. For example, a 20 N vector should be twice as long as a 10 N vector (Figure 1.1). An exact scale for the magnitude is not always required but it is important that vectors are drawn relative to one another.

20 N 10 N

FIGURE 1.1 An arrow representing a 20 N vector is twice as long as a 10 N vector.

Vector addition and subtraction

When more than one vector acts on an object, the overall or combined effect of the vectors can be found. This is called vector addition. Combining vectors is also known as adding vectors. There are two methods for adding vectors in one dimension: graphically and algebraically.

Adding vectors graphically

You can add vectors in one dimension graphically, using vector diagrams. After adding the vectors head-to-tail, the resultant vector can be drawn from the tail of the first vector to the head of the last vector (Figure 1.2). This is called the head-to-tail method.

9N + 4N = 13N

FIGURE 1.2 Vector addition.

Adding vectors algebraically

Vectors in one dimension can be added algebraically by applying a sign convention. Each vector is given a positive or negative sign, depending on its direction. It is up to you to decide which direction is positive. Then simply add up the vectors. The sign of the total tells you the direction of the vector.

For example, if Sally walked 2 metres west and then 5 metres east, either west or east could be the positive direction. If you choose east as the positive direction, then Sally walked negative 2 metres west and then positive 5 metres east. So her resultant displacement would be $(-2) + 5 = +3$ metres, which is 3 metres east.

Subtracting vectors

Vector subtraction is similar to vector addition. To find the difference between (or change in) vectors, subtract the initial vector from the final vector, remembering to apply an appropriate sign convention. The Greek delta symbol Δ, which means a difference or change, is used to indicate the resultant vector.

For example, to find the difference between a velocity of $50\,m\,s^{-1}$ south and a velocity of $20\,m\,s^{-1}$ north algebraically, subtract the initial velocity ($50\,m\,s^{-1}$ south) from the final vector ($20\,m\,s^{-1}$ north). If you have taken north to be the positive direction, then the resultant velocity $\Delta\vec{v} = +20 - (-50) = +70\,m\,s^{-1}$, that is, $70\,m\,s^{-1}$ north. This is shown graphically in Figure 1.3.

Original: $50\,m\,s^{-1}$ $20\,m\,s^{-1}$

To find the difference: $50\,m\,s^{-1}$ $20\,m\,s^{-1}$ $70\,m\,s^{-1}$

FIGURE 1.3 Vector subtraction.

ISBN 978 1 4886 1935 9 Pearson Physics 11 NSW | Skills and Assessment | Module 1 3

Worksheets

A diverse offering of instructive and self-contained worksheets is included in each module. Common to all modules are the initial 'Knowledge review' worksheet to activate prior knowledge, a 'Literacy review' worksheet to explicitly build understanding and application of scientific terminology, and finally a 'Thinking about my learning' worksheet, which students can use for reflection and self-assessment. Other worksheet types provide opportunities to revise, consolidate and further student understanding.

All worksheets function as formative assessment and are clearly aligned to the syllabus. A range of questions building from foundation to challenging are included within worksheets.

Practical activities

Practical activities give students the opportunity to complete practical work related to the various themes covered in the syllabus. All practical activities referenced in outcomes within the syllabus have been covered. Across the suite of practical activities, students have opportunities to design, conduct, evaluate, gather and analyse data, appropriately record results and prepare evidence-based conclusions. Students have opportunities to evaluate safety and risk, and identify any potential hazards.

Each practical activity includes a suggested duration. Along with the depth studies, the practical activities meet the 35 hours of practical work mandated at Year 11 in the syllabus. Where there is key knowledge that will support the completion of a practical activity, students are referred back to it.

Like the worksheets, the practical activities include a range of questions, building from foundation to challenging.

Depth study

Each module contains at least one suggested depth study. The depth studies allow further development of one or more concepts found within or inspired by the syllabus. They allow students to acquire a depth of understanding and take responsibility for their own learning, and promote differentiation and engagement.

Each depth study allows for the demonstration of a range of Working scientifically skills, with all depth studies addressing the Working scientifically outcomes of Questioning and predicting and Communicating. A minimum of two additional Working scientifically skills and at least one Knowledge and Understanding outcome are also assessed.

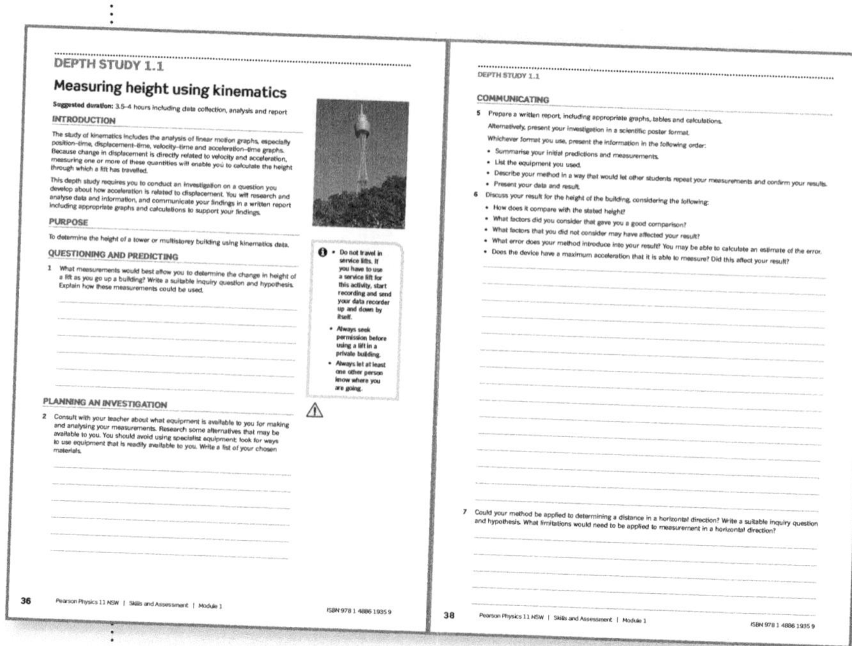
DEPTH STUDY 1.1

Measuring height using kinematics

ISBN 978 1 4886 1935 9

Module review questions

Each module finishes with a comprehensive set of questions, consisting of multiple choice, short answer and extended response, which helps students to draw together their knowledge and understanding and apply it to these styles of question.

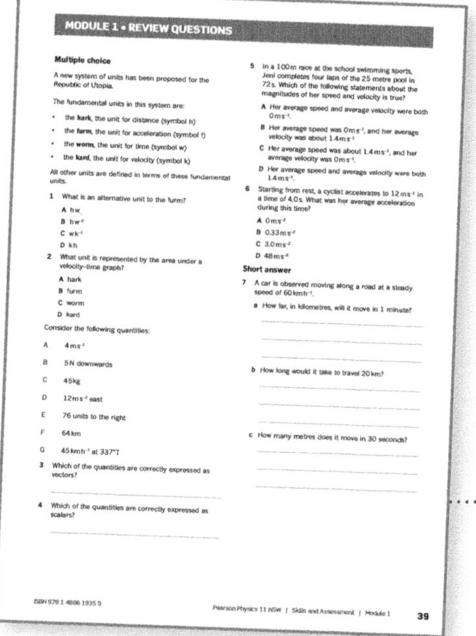

MODULE 1 • REVIEW QUESTIONS

Multiple choice

A new system of units has been proposed for the Republic of Utopia.

The fundamental units in this system are:

- the **hark**, the unit for distance (symbol h)
- the **furm**, the unit for acceleration (symbol f)
- the **worm**, the unit for time (symbol w)
- the **kard**, the unit for velocity (symbol k)

All other units are defined in terms of these fundamental units.

1 What is an alternative unit to the furm?

A hw

B hw^{-2}

C wk^{-1}

D kh

2 What unit is represented by the area under a velocity–time graph?

A hark

B furm

C worm

D kard

Consider the following quantities:

A $4\,m\,s^{-1}$

B 5 N downwards

C 45 kg

D $12\,m\,s^{-2}$ east

E 76 units to the right

F 64 km

G $45\,km\,h^{-1}$ at 337°T

3 Which of the quantities are correctly expressed as vectors?

4 Which of the quantities are correctly expressed as scalars?

5 In a 100 m race at the school swimming sports, Jeni completes four laps of the 25 metre pool in 72 s. Which of the following statements about the magnitudes of her speed and velocity is true?

A Her average speed and average velocity were both $0\,m\,s^{-1}$.

B Her average speed was $0\,m\,s^{-1}$, and her average velocity was about $1.4\,m\,s^{-1}$

C Her average speed was about $1.4\,m\,s^{-1}$, and her average velocity was $0\,m\,s^{-1}$.

D Her average speed and average velocity were both $1.4\,m\,s^{-1}$.

6 Starting from rest, a cyclist accelerates to $12\,m\,s^{-1}$ in a time of 4.0 s. What was her average acceleration during this time?

A $0\,m\,s^{-2}$

B $0.33\,m\,s^{-2}$

C $3.0\,m\,s^{-2}$

D $48\,m\,s^{-2}$

Short answer

7 A car is observed moving along a road at a steady speed of $60\,km\,h^{-1}$.

a How far, in kilometres, will it move in 1 minute?

b How long would it take to travel 20 km?

c How many metres does it move in 30 seconds?

ISBN 978 1 4886 1935 9 Pearson Physics 11 NSW | Skills and Assessment | Module 1 39

Rating my learning

This feature is an innovative tool that appears at the bottom of the final page of most worksheets and all practical activities. It provides students with the opportunity for self-reflection and self-assessment. It encourages them to look ahead to how they can continue to improve, and it helps them to identify focus areas for further skill and knowledge development.

The teacher may choose to use student responses to the 'Rating my learning' feature as a formative assessment tool. At a glance, teachers can assess which topics and which students need intervention for improvement.

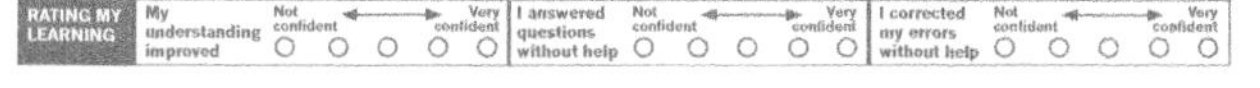

Icons and features

The 2018 New South Wales Physics Stage 6 Syllabus Learning Across the Curriculum content is addressed and identified.

The **safety icon** highlights significant hazards, indicating caution is needed.

The **safety glasses icon** highlights that protective eyewear is to be worn during the practical activity.

Highlight boxes focus students' attention on important information such as key definitions, formulae and summary points.

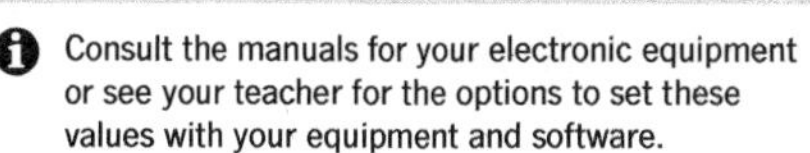

Teacher Support

Comprehensive answers and fully worked solutions for all worksheets, practical activities, depth studies and module review questions are provided via the *Pearson Physics 11 New South Wales* Teacher Support. An editable suggested assessment rubric for depth studies is also provided.

Pearson Physics 11 New South Wales

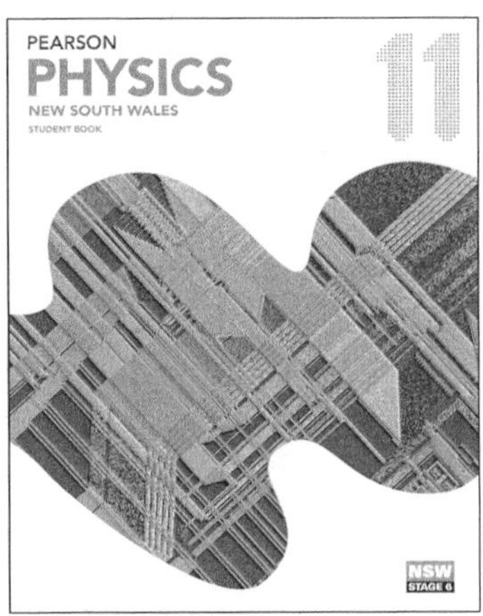

Student Book

Pearson Physics 11 New South Wales has been written to fully align with the new New South Wales Physics Stage 6 Syllabus. The Student Book includes the very latest developments and applications of physics and incorporates best-practice literacy and instructional design to ensure the content and concepts are fully accessible to all students.

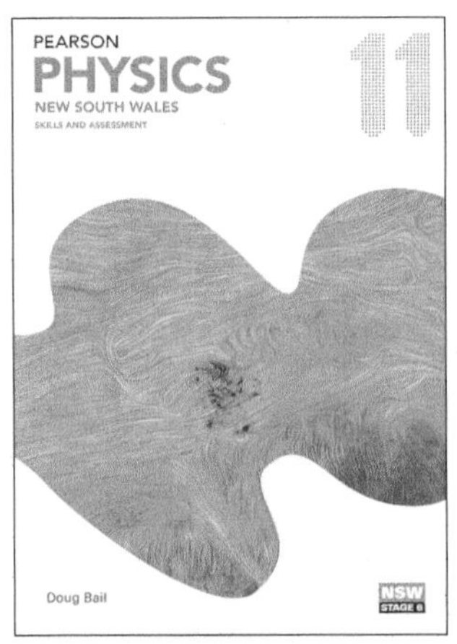

Skills and Assessment Book

The *Skills and Assessment Book* gives students the edge in preparing for all forms of assessment. Key features include a toolkit, key knowledge summaries, worksheets, practical activities, suggested depth studies and module review questions. It provides guidance, assessment practice and opportunities to develop key skills.

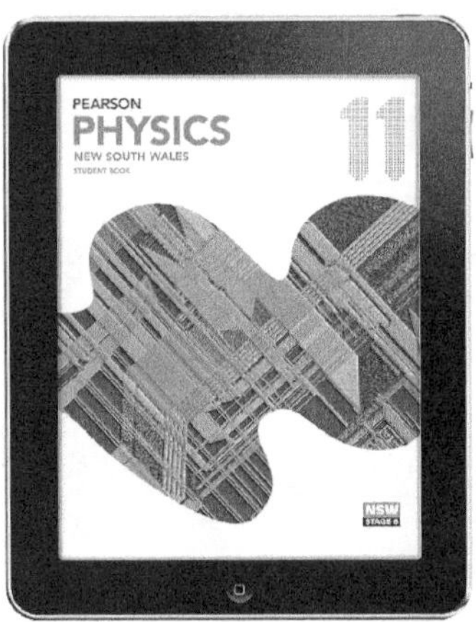

Reader+ the next generation eBook

Pearson Reader+ lets you use your *Student Book* online or offline on any device. Pearson Reader+ retains the look and integrity of the printed book. Practical activities, interactives and videos are available on Pearson Reader+ along with fully worked solutions to the Student Book questions.

Teacher Support

The Teacher Support available includes syllabus grids and a scope and sequence plan to support teachers with programming. It also provides the fully worked solutions and answers to all *Student Book* and *Skills and Assessment Book* questions, including all worksheets, practical activities, depth studies and module review questions. Teacher notes, safety notes, risk assessments and a laboratory technician's checklist and recipes are available for all practical activities. Depth studies are supported with suggested assessment rubrics and exemplar answers.

Pearson Digital

Access your digital resources at **pearsonplaces.com.au**
Browse and buy at **pearson.com.au**

ISBN 978 1 4886 1935 9

Physics toolkit

This toolkit provides support for developing the skills to complete the worksheets, undertake research and practical investigations and complete the depth studies. It covers examination techniques and study skills. The toolkit can serve as a reference tool to be consulted as needed throughout the year.

Practical investigation

A practical investigation or activity is a laboratory experiment and may involve fieldwork. Experiments are conducted to test a hypothesis by generating primary data that is analysed to draw a conclusion. You are required to complete a minimum of 35 hours of practical work during this course, which may include a depth study practical investigation.

CONDUCTING A PRACTICAL INVESTIGATION

Practical investigations follow a precise scientific procedure. The following checklist provides a summary of essential elements of a science practical investigation and the experiment report. Refer to the checklist as you conduct a student-designed practical investigation.

Elements	Explanation	Tick ✔
Date	The date the practical was conducted and the report was written.	
Title/heading	A statement of what is being investigated.	
Purpose/aim	• A statement that describes what the scientist wants to show, prove or find out in the experiment. • May be a statement or question. • May be one or two sentences. • Often written as 'To investigate the effect of X on Y.'	
Hypothesis	• An educated guess about the expected result of the experiment. • A testable prediction, often in the form of 'If X is true and this is tested, then Y will occur'.	
Variables	• Experiments measure relationships between variables and measures results. • Independent variable is the one that is changed. • Dependent variable is the one responding to the change of the independent variable. • Controlled variables are kept constant throughout the experiment.	
Risk assessments/ safety	• An analysis of the potential dangers of conducting the experiment. • Objective is to maximise safety by managing and minimising risk. • The risk assessment form on page xi guides you through the analysis of risk.	
Materials	• List of all equipment, chemicals and materials used. • Includes quantities (chemicals) and sizes (equipment).	
Procedure/method	• Detailed, numbered, step-by-step list of exactly what was done. • Includes quantities of chemicals used. • Can include 2D scientific drawings.	
Results	• A record of all observations and measurements taken during experiment. • Observations may be recorded as text, diagrams, photos, videos. • Most common records are presented as tables, graphs and can include calculations. • Check for errors and mistakes in data collected. Take measures to reduce these.	
Discussion	• Interpretation of results. • Includes analysis of the data, looking for trends and relationships. • Look for any problems such as bias, inaccuracy, unreliable data and errors. • Evaluate the method used, including a discussion of limitations. • Comment on whether results support or do not support the hypothesis.	
Conclusion	• Short statement based on the evidence that summarises the findings of the experiment. • Conclusion relates back to the purpose and the hypothesis, either supporting or not supporting the hypothesis. • Evidence should be clearly listed in support of the conclusion reached.	

RISK ASSESSMENT FORM

Five levels of safety should be considered in an investigation. The inverted pyramid ranks these levels in order of importance. The school and your teacher are responsible for reducing most of these risks. You as a student scientist can take measures to reduce the risks shown at the bottom of the hierarchy.

Complete the risk assessment form to identify possible risks for which you can take responsibility and to think of ways you can reduce risks to create a safe experiment environment.

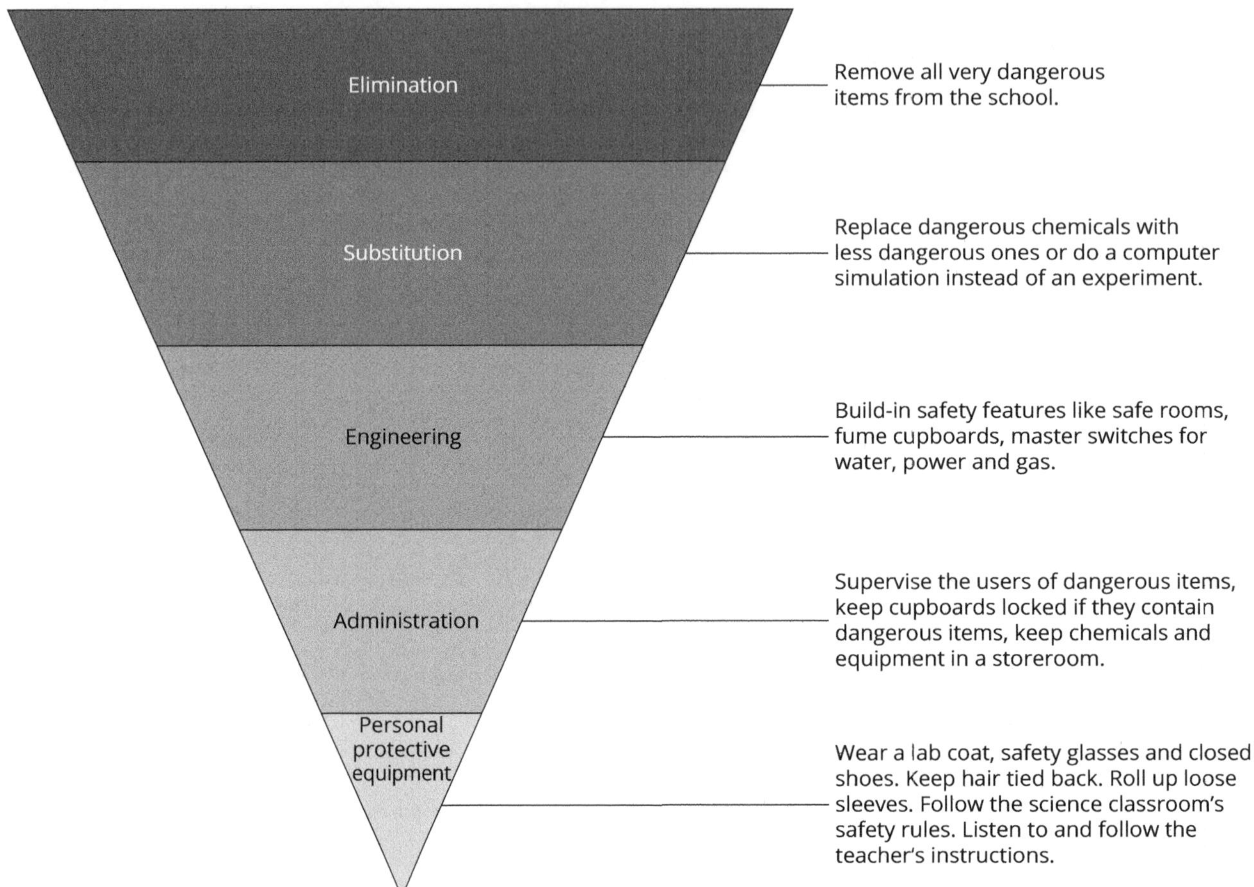

Risk assessment form		
What activity are you doing?		
Title or description of the investigation		
List:	**Identify any risks**	**State how you will:**
equipment you will be using		safely use each piece
chemicals or radioactive sources you will be using		carefully use each chemical or source carefully dispose of the chemicals
ethical issues you need to consider		ethically use animals in the laboratory ethically use human participants in the investigation
outdoor or fieldwork activities		reduce these risks
any other possible risks		reduce these risks

ISBN 978 1 4886 1935 9

EXAMPLES OF PRACTICAL REPORTS

It can be difficult to gauge whether you have attained a high standard in your completed practical investigation report. Looking at sample practical reports can help you identify what is required. Two sample practical reports are provided: one has room for improvement while the other is well done. The annotations draw your attention to key points to note on each practical report. These points are also reflected in the checklist, so you are able to use this as a tool to evaluate whether all requirements of the practical are complete.

High-standard practical report

Investigation of motion using timed intervals

Purpose

To investigate the motion of a student using timed intervals of measured distance.

Background

The timing of intervals during sprints, swimming competitions and racing is a key indicator to performance. Graphing position against time or motion against time provides a visual indicator of performance during stages of the event.

The hypothesis is based on the purpose of the investigation.

Hypothesis

If the position of an object is known at regular timed intervals, the velocity can be calculated using the equations of motion.

All materials are listed, including numbers of specific items.

Materials

- plastic cones or other position markers
- 30 metre tape measure
- stopwatches

Clear instructions are provided about each step of the experiment. These have been written in recipe style, with easy-to-follow, detailed instructions written in the third person.

Procedure

1 In a large, clear space stretch out the 30 metre measuring tape on the ground in a straight line. Mark out the beginning and end of the distance along with every 5 metre interval.

2 Station a student with a stopwatch at each 5 metre marker. Ensure each watch is zeroed and that students are familiar with the basic operation of the stopwatch.

3 Select a student for testing and position them ready to run, walk, crawl or hop from the start of the 30 metres. On a call of 'GO', the student starts to move along the marked distance. All student timers start their stopwatches.

4 As the student passes each timer, the timer stops his/her stopwatch and notes the time. Collate all times for this first trial.

5 Repeat the trial for 2 or 3 other students. Repeat any trials where timing students are incomplete or obviously wrong.

6 Use these results to construct a graph of position versus time either manually or using a calculator or computer. From the position–time graph, or otherwise, construct a graph of velocity versus time for a particular trial.

Data and analysis

Results have been checked for consistency between trials. The number of significant figures shown indicates a good understanding of the limitations of the measuring technique.

TABLE 1 Results of trials

Distance travelled (m)	Trial 1 Time taken (s)	Trial 2 Time taken (s)	Trial 3 Time taken (s)	Trial 4 Time taken (s)
0.0	0.0	0.0	0.0	0.0
5.0	2.3	2.2	2.3	2.4
10.0	4.2	4.1	4.2	4.3
15.0	6.5	6.6	6.4	6.5
20.0	8.4	8.4	8.3	8.4
25.0	10.3	10.2	10.4	10.3
30.0	14.2	14.1	14.3	14.2

ISBN 978 1 4886 1935 9

1 Graph position versus time for 2 or 3 trials and compare the student's performance.

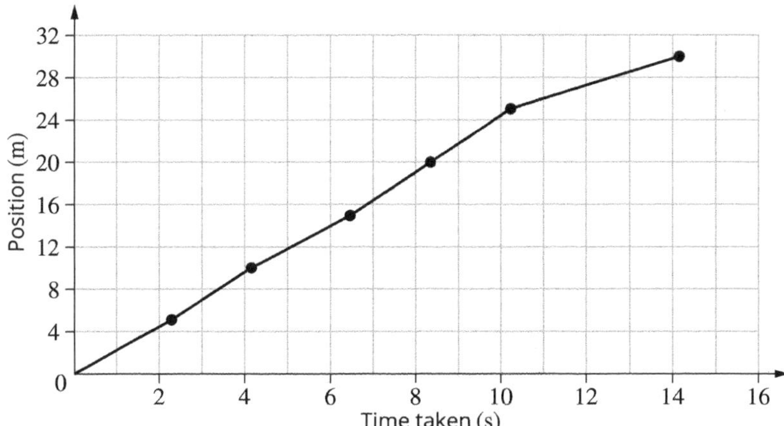

FIGURE 1 Position–time graph

2 From the position versus time graph, or otherwise, construct a velocity versus time graph for one particular trial.

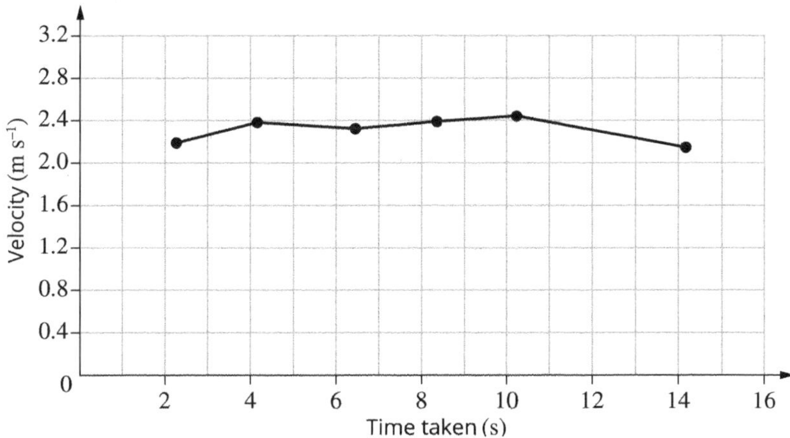

FIGURE 2 Velocity–time graph

Labelling of axes is complete and the graph includes a suitable descriptive title. An appropriate scale has been chosen that best utilises the space available.

3 Describe the means by which you found velocity from the position versus time graph.

The gradient of a position–time graph can be used to calculate velocity since the change in position is displacement. Alternatively, the individual 5 metre intervals can be used to calculate the velocity for each interval, i.e. $\vec{v} = \frac{\Delta \vec{s}}{\Delta t} = \frac{5}{t}$

Clear explanation with examples of how the calculation can be done. The response has been written in the third person.

Conclusion

4 Comment on the reliability of this means of measuring position, time and, hence, velocity.

The error in position can be expected to be small using a 30 metre tape to mark 5 metre intervals. Hand-held timing and watching students pass each marker introduces errors that can be significant if the student's velocity is large.

Explanation shows good understanding of the method and consideration of real sources of error. Avoid general comments unrelated to the specific investigation.

5 What is the major source of error in this activity?

Timing errors are the largest. A reaction time of 0.2 s is a large part of the time per interval.

Comments are short but precise. Shows a good understanding of the method. There is no need to write a long response when a short, well written response provides a clear answer.

6 Suggest an alternative that would improve the reliability of results.

Electronic timing methods such as a timing gate as used in electronics would eliminate errors associated with hand-held timing.

Response clearly addresses the sources of error identified in the previous response.

7 A ticker timer uses a constant time interval rather than a constant distance interval. Would this change the procedure by which you could calculate velocity? Which procedure would you expect to give the best results?

The ticker timer provides a more accurate timing method at the expense of errors due to friction between paper tape and timer.

The response has addressed both the positives and negatives. It could be extended to suggest a means of improving the ticker timer or considering the comparative errors in each method. A calculation of indicative percentage error in each method would be a useful addition.

Low-standard practical report

No table title.

Check over data carefully to be sure that it is entered in the appropriate row and column, Data has clearly been entered incorrectly in trial 1.

The number of significant figures shown is greater than the precision the measuring technique allows.

Investigation of motion using timed intervals

Purpose

To investigate the motion of a student using timed intervals of measured distance.

Background

The timing of intervals during sprints, swimming competitions and racing is a key indicator of performance. Graphing position against time or motion against time provides a visual indicator of performance during stages of the event.

Materials

- plastic cones or other position markers
- 30 metre tape measure
- stopwatches

Procedure

1 In a large, clear space stretch out the 30 metre measuring tape on the ground in a straight line. Mark out the beginning and end of the distance along with every 5 metre interval.

2 Station a student with a stopwatch at each 5 metre marker. Ensure each watch is zeroed and that students are familiar with the basic operation of the stopwatch.

3 Select a student for testing and position them ready to run, walk, crawl or hop from the start of the 30 metres. On a call of 'GO', the student starts to move along the marked distance. All student timers start their stopwatches.

4 As the student passes each timer, the timer stops his/her stopwatch and notes the time. Collate all times for this first trial.

5 Repeat the trial for 2 or 3 other students. Repeat any trials where timing students are incomplete or obviously wrong.

6 Use these results to construct a graph of position versus time either manually or using a calculator or computer. From the position–time graph, or otherwise, construct a graph of velocity versus time for a particular trial.

Data and analysis

Distance travelled (m)	Trial 1 Time taken (s)	Trial 2 Time taken (s)	Trial 3 Time taken (s)	Trial 4 Time taken (s)
0	1.5	0	0	0.000
5	2.3	2.2	2.3	2.356
10	4.2	4.1	4.2	4.323
15	6.5	6.6	6.4	6.523
20	6.5	8.4	8.3	8.421
25	10.3	10.2	10.4	10.284
30	14.2	14.1	14.3	14.232

ISBN 978 1 4886 1935 9

1 Graph position versus time for 2 or 3 trials and compare the student's performance.

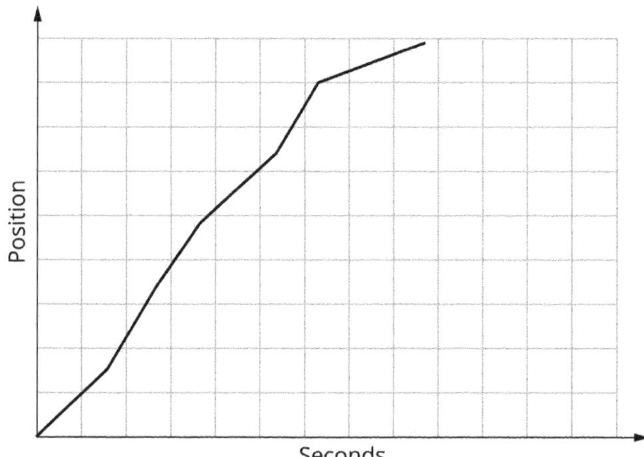

Graph should have a descriptive title and include a scale that utilises the graph area available

Labelling of axes is incomplete. Include both what is being graphed and the respective units. Scale should be clearly shown.

2 From the position versus time graph, or otherwise, construct a velocity versus time graph for one particular trial.

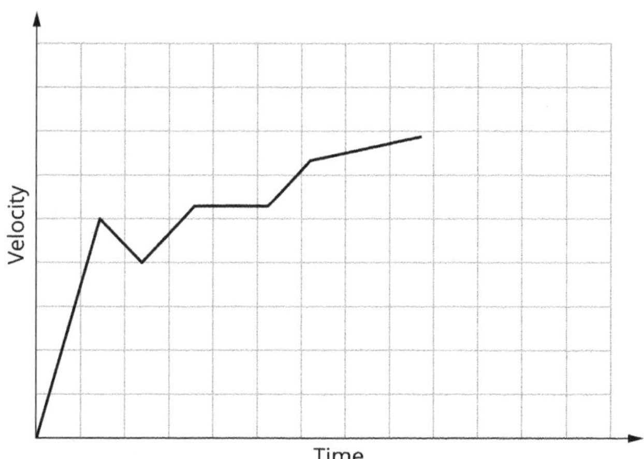

Incomplete labelling of graph. Needs title, axes both labelled with quantities and units as well as appropriate scale.

3 Describe the means by which you found velocity from the position versus time graph.

I used the gradient of the graph.

Explanation of steps involved is incomplete. Answers should be in the third-person, passive voice, e.g., 'The gradient of the graph was used...'

Conclusion

4 Comment on the reliability of this means of measuring position, time and, hence, velocity.

There were some big errors because the answers didn't match. This meant the velocity was wrong.

Explanation incomplete and not related back to actual data. Should be written in third-person, passive voice.

5 What is the major source of error in this activity?

I think the biggest errors came from not reading the instructions. Also the timing was hard and not everybody tried.

Comments should relate to the measurements and include direct reference to them. Always write in third-person, passive voice.

6 Suggest alternatives that would improve the reliability of results.

You could try measuring it again or maybe try a different method.

7 A ticker timer uses a constant time interval rather than a constant distance interval. Would this change the Procedure by which you could calculate velocity? Which Procedure would you expect to give the best results?

The procedure that gives the best results is the ticker timer because it measures 50 times per second. If you measure distance it changes so it won't be the same.

Take care to analyse exactly what was done and how. The response suggests poor understanding of the method. Always write in third-person, passive voice.

Secondary-sourced investigation

This section guides you in conducting a secondary-sourced investigation. For assistance with conducting a practical investigation, see pages ix–xv.

A secondary-sourced investigation is also often known as a research project. Such investigations require that you think carefully about the topic, find, collect and organise information, analyse and synthesise findings and present your ideas. The investigation process is summarised in the following flow chart. An investigation is not necessarily a straightforward linear process as shown in the flow chart. You can move back and forth between steps as needed.

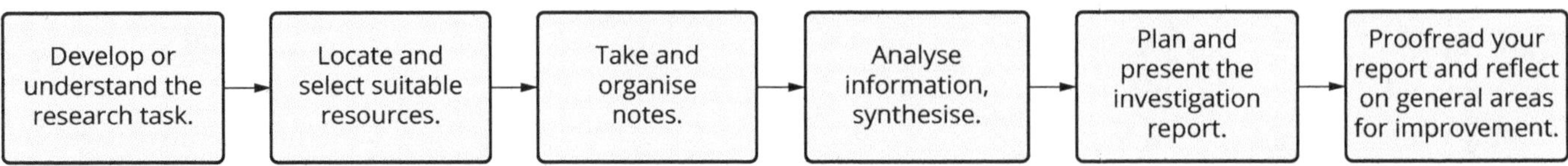

THE INVESTIGATION TASK

It is important to understand the breakdown of investigation questions or tasks. This helps you to write your own tasks and better understand the requirements of those written by others.

As you develop an investigation task, be aware of the depth of thinking it will require. The following chart provides support in writing tasks at different levels of thinking or complexity. It also provides some key words to help target tasks at these thinking levels and gives some examples of questions.

When developing your question or task, be conscious of the level at which you are pitching it. When you are writing your own questions for a depth study investigation, they should generally be at the analysis level.

Level of question complexity	Type of thinking	Words that may be used		Examples of questions
Simple	Retrieval of information—remembering, producing information on demand	• who • where • list • show • describe • select • complete • define	• what • when • label • demonstrate • name • state • recognise	**1** Define the term 'momentum'. **2** What factors affect inertia in a collision? **3** List the vector quantities in straight-line motion. **4** What variables affect the movement of thermal energy between two objects?
↓	Comprehension—the ability to understand information	• who • where • explain • represent • show how • represent	• what • when • summarise • draw • describe	**1** Represent the process of energy transfer with a well-labelled diagram. **2** Why are houses wired as a parallel circuit instead of a series circuit? **3** Explain why heat conductors feel cold to the touch. **4** What happens to a metal rod when it is heated?
↓	Analysis of information—scrutinising and breaking something into its smaller parts, including: • comparing • classifying • identifying errors • concluding • predicting • judging	• why • categorise • contrast • sort • organise • generalise • evaluate • edit • assess • judge	• how • compare • distinguish • discriminate between • deduce • critique • diagnose • identify errors • identify misunderstandings	**1** Organise the following in order from the simplest measure to highest level: velocity, acceleration, displacement. **2** Compare a solenoid with an electromagnet. **3** Compare and contrast alpha and beta decay.
Complex (requiring more thinking)	Application of information—using knowledge in new situations, including: • testing a hypothesis • solving a problem • experimenting and using data • decision-making.	• why • investigate • find out about • test • solve • develop • decide	• how • research • experiment • predict • adapt • judge	**1** Early scientists believed that heat energy was transferred by the spilling of 'caloric'. Construct an argument for or against this statement. **2** Investigate the behaviour of a silicon diode in a simple circuit. **3** How would conduction of heat energy in an iron rod differ at different atmospheric temperatures?

 ISBN 978 1 4886 1935 9

RESOURCES

The resources you refer to in an investigation may be primary or secondary, or both.

Primary sources of information are from investigations that you have conducted yourself. Examples of primary sources are results from your experiments, reports of your scientific investigations, photographs you have taken, and specimens or artefacts that you have collected. Secondary sources of information are from investigations that have been conducted by others. Secondary sources include peer-reviewed articles, textbooks, biographies, documentaries, newspaper articles, and many websites.

Refer to the two checklists below. The first shows features to look for when assessing and selecting the best sources for your investigation. The second shows the information that is required for the references section of your report.

Always remember to note the reference information for every resource you use. It is very time-consuming and difficult to backtrack to obtain these details when you need them to write the references.

Selecting resources for the investigation	Tick ✔
The resource is:	
• **credible** and I can identify the author, author's expertise and publisher	
• **current** because the date of publication of the material is provided and is recent	
• **factual** and I know that it is objective material and not biased	
• **accurate** and all information is correct	
• **relevant** and covers the area I am investigating	
• **readable** and neither too simple nor too complex in its coverage of the material	

Information for references and examples
Article in scientific journal or magazine Author(s), initials (year). Title of article. *Journal title, volume number*(issue number). page numbers. Kensrud, J.R., Nathan A.M. and Smith L.V. (2017). Oblique collisions of baseballs and softballs with a bat. *American Journal of Physics 85*(8). 503–509.
Book Author(s), initials (year). *Title of book* (edition, if not first). City: Publisher. Rickard, G. *et al.* (2017), *Pearson Science Student Book 9*. Melbourne: Pearson Education. If there are more than three authors, list first author then write et al. (in italics), meaning 'and others'.
Internet Author(s) or name of organisation (year). *Title of web page or web document.* Retrieved from <URL>. American Institute of Physics (2017). *Periodic table.* Retrieved from http://history.aip.org/history/exhibits/electron/ Date accessed

NOTE-TAKING AND ORGANISING NOTES

Note-taking and organising notes takes skill. Good note-taking helps you avoid plagiarism and provides excellent information for writing up the investigation. Plagiarism is taking someone else's ideas and words and presenting them as your own work. You plagiarise if you copy sections or sentences from sources, or if you cut and paste from the internet. It is acceptable to use the ideas of others, but you must state clearly where the information has come from in the body of your report and include the source in your references.

The examples in the following table show the original text, plagiarised text, and acceptably rephrased text.

Original text	Plagiarised text	Rephrased text
Dogs have sweat glands on their feet. Dogs pant when they are hot because their sweat glands are not sufficient to cool down their bodies. In addition, their tongues allow the water from their bodies to evaporate and cool down their bodies.	Dogs have their sweat glands on their feet. When dogs get hot they pant because their sweat glands are not enough to cool down their bodies. Their tongues let the water from their bodies evaporate and cool their bodies.	Dogs pant in order to cool down. Water evaporates from their tongues and this lowers their body temperature. They can't get cool enough through just the sweat glands on their feet.

There are various approaches to effective note-taking. Whatever technique you use, try to keep notes brief, and focus on key points. Some examples include:

- dot point summary
- underlining or highlighting text
- labelled diagram
- flow chart: show sequences

- concept map: show connections between items
- Venn diagrams: show similarities and differences
- table: may incorporate any of the other note-taking techniques. Tables are useful for summarising longer and more complex information that has subparts. Adapt the table to suit your style and the task. The following sample table (which is partially completed) shows how this technique can be used to take notes for your secondary-sourced investigation.

Secondary-sourced investigation task Many scientists believe that limiting human population growth is necessary to control environmental damage. Construct an argument for or against this statement.			
	Source 1 (e.g. book) Title: Author: Publisher's name: Publisher's location: Date of publication:	**Source 2** (e.g. internet) Title: Author: URL: Date accessed:	**Source 3** (e.g. science journal) Author: Date: Title of article: Journal title: Volume number: Pages:
Population growth trends		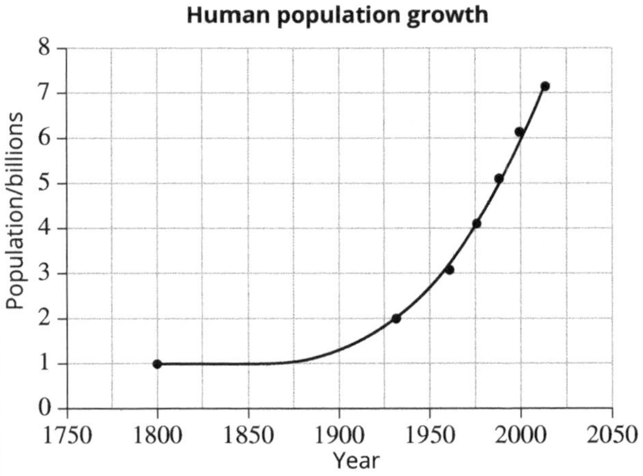 **Human population growth** Population/billions: 0, 1, 2, 3, 4, 5, 6, 7, 8 Year: 1750, 1800, 1850, 1900, 1950, 2000, 2050	
Impact of population growth on environment	• growth of cities • demand for resources		
Reasons to control and limit population growth	• increased demand for resources for human survival		
Reasons not to limit population growth but allow natural population growth			• other factors contribute to environment damage; land-use policies, e.g. poor land use

SCIENTIFIC WRITING

Scientists have a particular writing style. Your investigation should use this distinctive style to communicate your ideas. Writing in the scientific style is:

- objective—describes events rather than what people think or feel
- free of bias or personal opinion
- precise—avoids exaggeration and uses qualified language
- formal—scholarly language rather than colloquial or everyday language
- concise—conveys information in short, clearly understandable sentences without unnecessary information
- simple—short sentences used where possible
- usually written in passive voice rather than active voice, although the active voice can be used sometimes
- structured to include headings, tables, diagrams, mathematical calculations.

ISBN 978 1 4886 1935 9

Examples of unscientific and scientific writing are demonstrated in the table below.

Unscientific writing	Scientific writing
Subjective, biased writing • The results were fantastic. • This produced a disgusting odour. • The breathtakingly beautiful bowerbird...	**Objective, unbiased writing** • The results showed... • This produced a pungent odour ... • The golden bowerbird...
Exaggerated writing • The object weighed a huge amount. • The magnesium burst into huge flames. • Millions of ants swarmed over...	**Accurate, precise writing** • The mass of the object was 250 kg. • The magnesium burnt vigorously. • Ants swarmed all over...
Everyday, informal language • The bacteria passed away. • The results don't... • We guessed that... • Previous researchers were slack and missed...	**Formal language** • The bacteria died. • The results do not... • It was hypothesised that... • Previous researchers did not perceive that...
Active voice • We recorded oxygen levels every hour. • We put 50 g of solute in a conical flask containing distilled water, and then we slowly added 1 mol L^{-1} hydrochloric acid.	**Passive voice** • The oxygen level was recorded hourly. • 50 g of solute was placed in a conical flask containing distilled water, and then 1 mol L^{-1} of hydrochloric acid was slowly added.

PRESENTING THE INVESTIGATION

Scientific findings may be presented in a variety of ways. A common presentation format at science conferences is a poster. Posters can get ideas across to a large audience in an organised, concise and creative way. Other common presentation formats are essays, reports, orals or articles. Each presentation format has its own conventions. The table below summarises characteristics of a number of presentation formats.

Presentation formats and their characteristics		
Format	**Characteristics/inclusions**	
Poster	• balance of text and visuals • title, subheadings • balanced layout • captions for figures and tables	• references • hierarchy of font size according to subheading level • consistent font style: no more than three fonts
Report/article	• structured with introduction, paragraphs, conclusion • includes subheadings	• mainly text • can include diagrams, graphs, tables
Essay	• structured with an introduction, paragraphs, conclusion • introduction states focus of essay • each paragraph makes a new point supported by evidence	• each paragraph links back to last paragraph • a text-style presentation format – visuals at end in appendix • conclusion draws all ideas together but does not include any new information
Oral	• needs to be engaging • use cue cards but do not read from them • watch audience as you speak	• stand still and don't fidget • look at audience and appear confident

PROOFREADING

After you have completed the investigation and prepared your presentation, it is important to think about and check what you have done.

Proofread your work to minimise errors and maximise communication of the ideas from your investigation. Use these questions as a proofreading checklist.

Proofreading checklist	Tick ✔
Have I investigated the question fully?	
Have I expressed myself clearly so I communicated my ideas well?	
Have I used the scientific writing style?	
Have I included data analysis?	
Have I checked spelling, punctuation and grammar?	
Have I included references?	
Have I met the requirements of the presentation format?	

Depth study

A depth study is an investigation that allows you to look in more detail into a particular area of interest in the syllabus. The depth study gives you the chance to gain greater understanding of concepts and is designed so you take more responsibility for your own learning.

It is expected that you demonstrate the use of a variety of scientific skills. Depth studies may vary, and may include practical activities, field work reports, research assignments, and may be based on primary or secondary data or sources. To fulfil the minimum 15 hour time requirement, your teacher may ask you to complete one very detailed depth study or a number of smaller depth studies.

Your teacher may provide a scaffolded approach with questions that guide you through a depth study, or you may be asked to design it yourself. This section of the toolkit will help you with planning your own depth study.

PLANNING THE DEPTH STUDY

Careful planning of the depth study before actually starting the investigation will help you to be clear about what you will do and how you will do it. Think of yourself as taking on the role of your teacher; identifying the depth study investigation topic, deciding how it will be investigated and presented, managing time through the investigation and checking that all syllabus requirements are met.

Clarify your ideas for your depth study, following these steps.

1 List all areas that interest you for your depth study. For example, forces in a collision or standing waves in a glass.	**My areas of interest are:** __________ __________ __________ __________
2 Do some quick research into each area of interest. Identify which gives you scope to develop into a depth study.	**My chosen area of investigation is:** __________ __________
3 Decide on the procedure you will use to conduct the depth study: • a practical investigation: - may begin with initial observations of a day to day activity - may be a secondary-sourced investigation - must include data analysis. • a secondary-sourced investigation: - may be expository (explain something) - may be an argument based on evidence - must include data analysis. • creating: - may design and construct a working model - may create a portfolio.	**My procedure will be:** __________ __________ __________ __________ __________ __________ __________

4 Decide how you will present the depth study. Select a presentation format that suits your type of investigation.
• a practical investigation
  - practical report
• a secondary-sourced (research) investigation
  - documentary
  - media report
  - literature review
  - visual presentation
  - journal article
  - essay
  - scientific poster
• designing or inventing or creating
• portfolio
• working model.
	I will present my depth study by:
	____________________
	____________________
	____________________
	____________________

5 Now that you have thought about what and how you will conduct the depth study, think about how you can best use your time. It is easy to spend most of the time conducting the practical or finding resources for the secondary-sourced investigation. This may leave little time to analyse data and prepare your presentation. Refer to the suggested planning timeline. Plan the depth study. — Time allocation 10% Conduct the investigation. — 40% Organise information and analyse data. — 30% Present information and proofread work. — 20%	**My depth study begins on** __________ **and is to be completed by** __________**.** Fill in the timeline to plan how you will allocate your time to complete the depth study, including specific dates.

6 Assessment requirements
Depth study must:
• address Working scientifically skills outcomes: Questioning and predicting, and Communicating
• address a minimum of two additional skills outcomes:
  - Planning investigations
  - Conducting investigations
  - Processing data and information
  - Analysing data and information
  - Problem solving
• must include at least one knowledge and understanding outcome.
	My depth study will include these Working scientifically skills:
	1 Questioning and predicting
	2 Communicating
	3 ____________________
	4 ____________________
	My depth study covers this/these knowledge and understanding outcome(s):
	____________________
	____________________

Study skills

There are a variety of techniques or strategies to help you study. You may find that you use different strategies in different situations. For example, you may prefer to highlight key phrases in your notebook throughout the year but make summaries of topics before an examination. The strategies you choose depend on personal preference and may not be the same as those used by classmates.

Effective study skills involve more than the learning strategies you use. Equally important is when you use skills. It is more effective to apply study skills throughout the year, revising and consolidating your knowledge as you progress through the course, rather than doing a rushed cram just before the examination. Revise work regularly. Being organised is the key to reducing your stress and setting up a study plan.

GETTING ORGANISED

To get yourself organised, try the following ideas:

- Use a diary to write down all homework and assessment tasks as soon as you get them. Note due dates and what you have to do.
- Be specific about the tasks you need to do. Rather than writing 'Do Physics' it is better to state things like which questions to answer and which page to look at in your student book.
- Write a list of everything you need to do each day. Tick off or cross out items as you complete them.
- Break down larger tasks into smaller separate parts that are manageable.
- Make sure your lists and planners are realistic. Do not set yourself more than you can actually do.

STUDY TECHNIQUES

Studying requires concentration. Remove any distractions and factor in some breaks. Have a break of 10 minutes every hour. Vary your study technique depending on the content to be learned and your personal preference. Although you may have already found a study technique that works for you, also consider the following options.

Study technique	Tips
Highlighting Radiation Radiation is a shortened form of electromagnetic radiation, which includes visible, ultraviolet and infrared light. Together with other forms of light, these make up the electromagnetic spectrum.The transfer of heat from one place to another without the movement of particles is by electromagnetic radiation. Electromagnetic radiation travels at the speed of light.	• Highlight or underline key points as you read your notes or text.
Summary notes *IONIC COMPOUNDS* • *Common properties of ionic compounds:* • *brittle, which means they shatter when hit with a hammer* • *hard, which means they are resistant to scratching* • *high melting point* • *unable to conduct electricity in solid state* • *good electrical conductor in liquid or aqueous states*	• Create a list of key headings and add some dot points about each heading. • Write your own summary of the key ideas in each chapter. • Use headings and subheadings. • Underline key words and key phrases. • Use simple diagrams. • The most effective chapter summaries are clear, to the point and uncluttered.
Diagrams 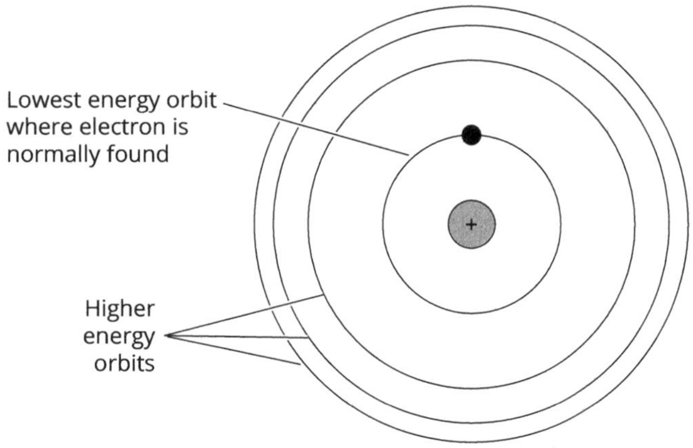 Lowest energy orbit where electron is normally found + Higher energy orbits	• Diagrams can be used as a summary of key concepts. • They are useful memory triggers. • Diagrams cover a lot of information in a visual way, with minimal text.
Concept maps 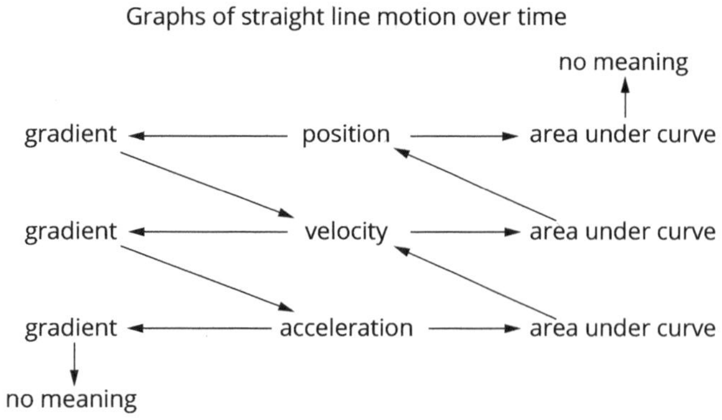 Graphs of straight line motion over time no meaning gradient ← position → area under curve gradient ← velocity → area under curve gradient ← acceleration → area under curve no meaning	• These are a great way of connecting key terms and ideas in a simple and ordered way. • Use the lines that connect ideas to note the relationships between the words and phrases. • They can include text and images. • They can be a simple or more complex summary tool. • Concept maps and other graphic organisers such as Venn diagrams and flow charts show how information is connected and help deepen understanding.

 ISBN 978 1 4886 1935 9

<table>
<tr><td>

Tables

<table>
<tr><th colspan="3">Subatomic particles</th></tr>
<tr><th>Relative mass</th><th>Charge</th><th>Location</th></tr>
<tr><td>1</td><td>neutral</td><td>nucleus</td></tr>
<tr><td>1</td><td>positive</td><td>nucleus</td></tr>
<tr><td>$\frac{1}{1836}$</td><td>negative</td><td>orbiting nucleus</td></tr>
</table>

</td><td>

- Tables are useful to show relationships between different factors.
- Information is uncluttered.

</td></tr>
<tr><td>

Mnemonic devices
Work **i**s **f**un, **s**ir.
...triggers your memory for...
$W = Fs$

</td><td>

- Mnemonic tools help you remember information.
- To remember that work is a product of force and displacement, memorise the sentence shown, which contains the initial letters of each variable.

</td></tr>
<tr><td>

Glossary
Electron: negatively charged particle in an atom
Diode: a semiconductor

</td><td>

- Compile your own glossary as writing the terms and definitions will make them easier to remember.
- Add images to help you remember.
- Write each term in a sentence.
- Memorise these terms and definitions.

</td></tr>
<tr><td>

Trigger words
Origin of universe– two theories:

- *Big Bang*
- *Steady state theory – Fred Hoyle*

</td><td>

- Write down the key words associated with a topic or theme.
- Trigger words are useful in helping to remember other related words and ideas.

</td></tr>
<tr><td>

Repeating information aloud

</td><td>

- This is a good way to remember and to force yourself to slow down and absorb the information.

</td></tr>
<tr><td>

Practising

</td><td>

- Do as many review questions and old exams as possible.

</td></tr>
<tr><td>

Flash cards

What is the Schrödinger model?

Identify the states of matter and changes between them.

</td><td>

- Making flash cards helps your understanding.
- Make your own cards.
- Write a question on one side and the answer on the back.
- Cards can include definitions, brief explanations, diagrams, equations, graphs.

</td></tr>
<tr><td>

Teaching someone

</td><td>

- Teach friends or family members.
- Teaching a difficult concept to someone means you must first understand the concept yourself.

</td></tr>
<tr><td>

Writing notes

</td><td>

- Hand-write rather than type summary notes.
- Remember the examination requires you to write answers.
- Practise writing for long stretches of time and make sure your writing is legible.

</td></tr>
<tr><td>

Responding to feedback and self-correcting

</td><td>

- Check through all feedback from your teacher.
- Highlight what was right or wrong.
- Attempt to identify where you have errors and rework the answer to get it right.

</td></tr>
</table>

EXAMINATION PREPARATION

In the weeks before the examination begin your exam preparation. The earlier you begin revising, the easier it will be. Now begin preparing for the examination.

Like most skills, practice will improve your ability to do exams and to handle different types of exam questions. Doing practice exams is vital because you gain experience in:

- using reading time effectively before starting to write
- allocating the right amount of time for each question
- working to a time limit
- reading and interpreting questions
- understanding what is required of each question
- planning answers
- deciding on relevant information
- proofreading/checking over your own answers
- writing efficiently for the duration of the exam.

Use the following checklist as a reminder of your study program.

Study program	Tick ✓
I have revised all areas of the course.	
I have highlighted important points.	
I have made a summary of the important points in each topic.	
I have read over my revision notes.	
I have looked at and worked through sample exam papers.	
I have answered practice questions in the appropriate time limit.	

EXAMINATION STRATEGIES

Familiarise yourself with the conditions of the examination well before the day you sit the exam. You should know:

- the number of exams for the subject
- the amount of time allowed for reading before the exam begins
- the amount of writing time allowed
- any particular equipment allowed and/or required, such as a calculator, lead pencils, pens, ruler etc.
- strategies to tackle the exam. Exam strategies are listed in the table that follows.

Exam strategies
Reading time • No writing at all is allowed during this time—no note-taking, no highlighting, no underlining. • Read the instructions. • Read through the short-answer questions first, followed by the extended-answer questions—this will give you an overall sense of the themes of questions that require written responses. • Read the multiple-choice questions next.
Writing time • Begin with the multiple-choice questions. • Answer every multiple-choice question, even if you can only make an educated guess. • If you are unsure of an answer to a multiple-choice question, mark it so you can come back to it if time allows. • Attempt the short-answer questions next. • Attempt the easiest short-answer questions first and work your way to the more challenging questions. • Attempt all other questions next.
Tips for answering questions • Carefully read each question, underlining key words. • Be aware that most questions are structured so they become more challenging towards the end. You may not be able to answer the last part of a question but you can earn most of the marks by answering the easier parts of the question. • Look carefully at any diagrams, pictures, tables and graphs and make sure you understand their relevance to the questions involved. • For questions with graphs, read the graph title and the labels on the axes carefully, so that you can establish the relationship the graph is showing. • For questions with tables, read the headings on the columns and rows carefully, so that you can analyse the content of the table effectively. • Check for the key words in a question. Highlight them but don't colour the whole question. • Plan your answers before you write, remembering to address the exam criteria. • For questions with parts, read the whole question first. This gives you an overall picture of the question. It will also help to ensure that you do not repeat yourself in subsequent parts of the question. • Make sure you actually answer the question that is asked. • Once you have answered the question, re-read your answer and then re-read the question, to ensure that you have actually answered the question. • When writing a definition, don't use the word you are defining in the actual definition. • If giving values from a graph, use a ruler to line up points with the axes so you can be accurate, and always include units in your answer. • Be sure to attempt all questions. • Read over your answers to pick up careless errors—the mind is faster than the hand, and you may not always write what you intend (especially when you have limited time). • Write legibly. Exams are scanned and marked online. If the assessor can't read your answer, they can't mark it as correct. • Keep an eye on the time. • *Never* leave an exam early. Use any spare time to re-read and check your answers.
Exam cues • The number of marks allocated to a question provides a clue about how much you are expected to write. Two marks usually means you need to make a minimum of two points. • The number of lines allowed for the answer indicates the length of the expected answer. If your writing is large, you may need to turn the page and continue on the back of the page. Make sure you indicate that the examiner must turn to the back of the page for the rest of the answer. Where the answer continues, clearly state that it is the continuation of the question and state the question number.

ISBN 978 1 4886 1935 9

MODULE 1 Kinematics

Outcomes

By the end of this module you will be able to:

- design and evaluate investigations in order to obtain primary and secondary data and information PH11-2
- conduct investigations to collect valid and reliable primary and secondary data and information PH11-3
- select and process appropriate qualitative and quantitative data and information using a range of appropriate media PH11-4
- analyse and evaluate primary and secondary data and information PH11-5
- solve scientific problems using primary and secondary data, critical thinking skills and scientific processes PH11-6
- describe and analyse motion in terms of scalar and vector quantities in two dimensions and make quantitative measurements and calculations for distance, displacement, speed, velocity and acceleration PH11-8

Content

MOTION IN A STRAIGHT LINE

INQUIRY QUESTION **How is the motion of an object moving in a straight line described and predicted?**

By the end of this module you will be able to:

- describe uniform straight-line (rectilinear) motion and uniformly accelerated motion through:
  - qualitative descriptions
  - the use of scalar and vector quantities (ACSPH060)
- conduct a practical investigation to gather data to facilitate the analysis of instantaneous and average velocity through: ICT
  - quantitative, first-hand measurements
  - the graphical representation and interpretation of data (ACSPH061) N
- calculate the relative velocity of two objects moving along the same line using vector analysis
- conduct practical investigations, selecting from a range of technologies, to record and analyse the motion of objects in a variety of situations in one dimension in order to measure or calculate: ICT N
  - time
  - distance
  - displacement
  - speed
  - velocity
  - acceleration

Module 1 • Kinematics

- use mathematical modelling and graphs, selected from a range of technologies, to analyse and derive relationships between time, distance, displacement, speed, velocity and acceleration in rectilinear motion, including:
  - $\vec{s} = \vec{u}t + \frac{1}{2}\vec{a}t^2$
  - $\vec{v} = \vec{u} + \vec{a}t$
  - $\vec{v}^2 = \vec{u}^2 + 2\vec{a}\vec{s}$ (ACSPH061) ICT N

MOTION ON A PLANE

INQUIRY QUESTION **How is the motion of an object that changes its direction of movement on a plane described?**

By the end of this module you will be able to:

- analyse vectors in one and two dimensions to:
  - resolve a vector into two perpendicular components
  - add two perpendicular vector components to obtain a single vector (ACSPH061) N
- represent the distance and displacement of objects moving on a horizontal plane using:
  - vector addition
  - resolution of components of vectors (ACSPH060) ICT N
- describe and analyse algebraically, graphically and with vector diagrams, the ways in which the motion of objects changes, including: ICT
  - velocity
  - displacement (ACSPH060, ACSPH061) N
- describe and analyse, using vector analysis, the relative positions and motions of one object relative to another object on a plane (ACSPH061)
- analyse the relative motion of objects in two dimensions in a variety of situations, for example:
  - a boat on a flowing river relative to the bank
  - two moving cars
  - an aeroplane in a crosswind relative to the ground (ACSPH060, ACSPH132) ICT N

Key knowledge

Motion in a straight line

In physics, the study of the motion of objects is called **kinematics**. Motion is described by physical quantities such as velocity, distance and time. These quantities can be divided into two broad groups: scalars and vectors.

SCALARS AND VECTORS

Scalars are quantities that have a magnitude but not a direction. An example of a scalar quantity is distance. When describing the distance of a race, the only thing needed is the numerical value and a unit, such as 400 metres.

Vectors are quantities that have a direction as well as a magnitude. An example of a vector quantity is force. The motion of an object resulting from an applied force depends on both the magnitude and direction of the force. For this reason, force is described by a magnitude, a direction and a unit, such as 6 newtons upwards. Vector quantities are written using vector notation. In this course this notation is an italic symbol with an arrow above it. For example, force is represented by $\vec{F}$.

TABLE 1.1 Examples of different scalar and vector quantities

Scalars	Vectors
time	force
distance	displacement
speed	velocity
energy	momentum

Vectors

When dealing with vectors, direction and magnitude must both be considered. A vector is represented visually by an arrow, and the length of the arrow represents the magnitude. For example, a 20 N vector should be twice as long as a 10 N vector (Figure 1.1). An exact scale for the magnitude is not always required but it is important that vectors are drawn relative to one another.

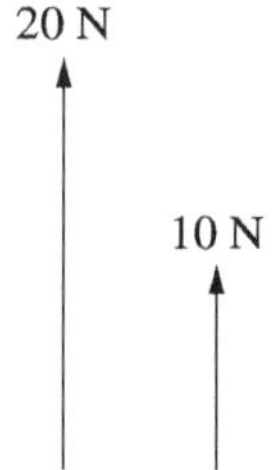

FIGURE 1.1 An arrow representing a 20 N vector is twice as long as a 10 N vector.

Vector addition and subtraction

When more than one vector acts on an object, the overall or combined effect of the vectors can be found. This is called vector addition. Combining vectors is also known as adding vectors. There are two methods for adding vectors in one dimension: graphically and algebraically.

Adding vectors graphically

You can add vectors in one dimension graphically, using vector diagrams. After adding the vectors head-to-tail, the resultant vector can be drawn from the tail of the first vector to the head of the last vector (Figure 1.2). This is called the head-to-tail method.

9N → + 4N → = 13N →

FIGURE 1.2 Vector addition.

Adding vectors algebraically

Vectors in one dimension can be added algebraically by applying a sign convention. Each vector is given a positive or negative sign, depending on its direction. It is up to you to decide which direction is positive. Then simply add up the vectors. The sign of the total tells you the direction of the vector.

For example, if Sally walked 2 metres west and then 5 metres east, either west or east could be the positive direction. If you choose east as the positive direction, then Sally walked negative 2 metres west and then positive 5 metres east. So her resultant displacement would be $(-2) + 5 = +3$ metres, which is 3 metres east.

Subtracting vectors

Vector subtraction is similar to vector addition. To find the difference between (or change in) vectors, subtract the initial vector from the final vector, remembering to apply an appropriate sign convention. The Greek delta symbol Δ, which means a difference or change, is used to indicate the resultant vector.

For example, to find the difference between a velocity of $50\,m\,s^{-1}$ south and a velocity of $20\,m\,s^{-1}$ north algebraically, subtract the initial velocity ($50\,m\,s^{-1}$ south) from the final vector ($20\,m\,s^{-1}$ north). If you have taken north to be the positive direction, then the resultant velocity $\Delta\vec{v} = +20 - (-50) = +70\,m\,s^{-1}$, that is, $70\,m\,s^{-1}$ north. This is shown graphically in Figure 1.3.

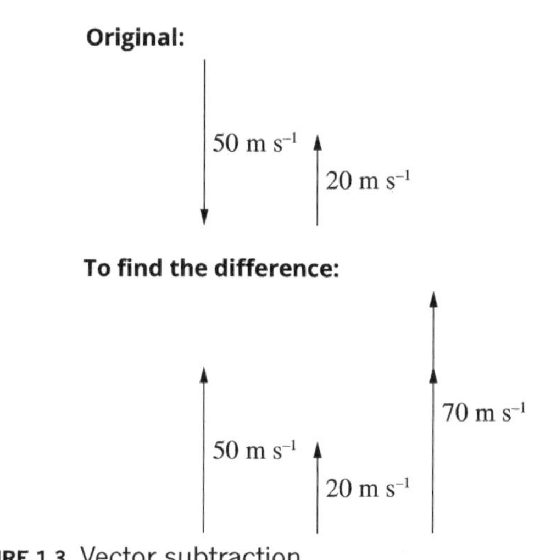

FIGURE 1.3 Vector subtraction

DISPLACEMENT, SPEED AND VELOCITY

In kinematics, the words used to describe motion have very precise meanings that must be understood clearly. Position, distance, displacement, speed and velocity are the focus in this section, because they are all used to describe **rectilinear motion**.

The motion of an object travelling in a straight line is known as rectilinear motion.

Position

Position defines the location of an object with respect to a defined origin. For example, you might see a bird take off from its nest in front of you and land on a post 100 m to your right. If the nest is considered to be the origin, its final position is 100 m to the right of the origin. It is important to note that the origin can be any reference point; it all depends on the information given, and what you are trying to determine.

Distance and displacement

Distance travelled, d, tells us how far an object has travelled in total, but not the direction. Distance travelled is therefore a scalar.

Displacement, $\vec{s}$, tells us the change in the position of an object, and the direction of the change. Displacement is therefore a vector: $\vec{s}$ = final position – initial position. If the final position is the same as the initial position, it is important to note a distance has been covered but the displacement is zero.

In Figure 1.4, a person practising for a race runs 100 m north to the finish line, then turns around and jogs south for 50 m. The distance they have travelled is therefore 150 m, and their displacement from the start is 50 m north.

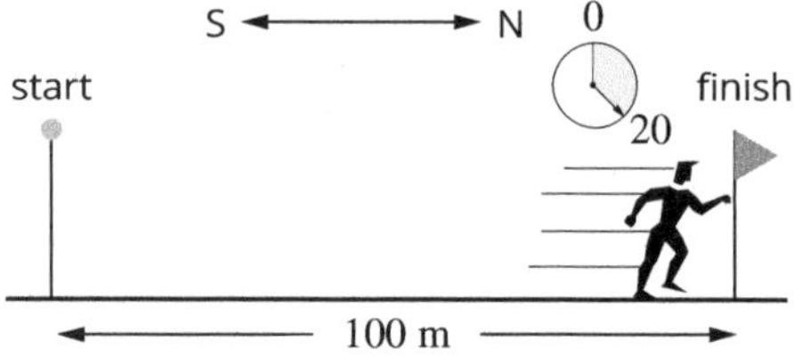

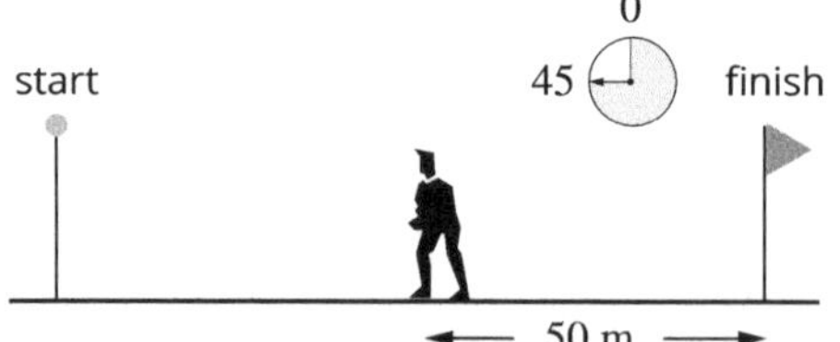

FIGURE 1.4 Distance and displacement

Speed and velocity

In everyday language 'speed' and 'velocity' are often used interchangeably, but in kinematics they are not the same. **Speed** is the rate of change of distance and is a scalar quantity represented by v. **Velocity** is the rate of change of displacement and is a vector quantity represented by $\vec{v}$. Instantaneous speed and velocity are measured at one point in time. Average speed and velocity are measured over a period of time.

The instantaneous speed of an object is a measure of how fast it is moving at a particular point in time. Its instantaneous velocity has the same magnitude as the instantaneous speed, but also has direction.

The formulae for average speed and average velocity are as follows:

$$\text{average speed } v_{av} = \frac{\text{distance travelled}}{\text{time taken}} = \frac{d}{\Delta t}$$

$$\text{average velocity } \vec{v}_{av} = \frac{\text{displacement}}{\text{time taken}} = \frac{\vec{s}}{\Delta t}$$

The SI unit for speed and velocity is metres per second (m s^{-1}), but kilometres per hour (km hr^{-1}) is often used when describing the motion of vehicles or aircraft. You can convert from one unit to the other by converting metres to kilometres and seconds to hours (or vice versa), but that can be difficult. Figure 1.5 shows a short cut for converting between these two units.

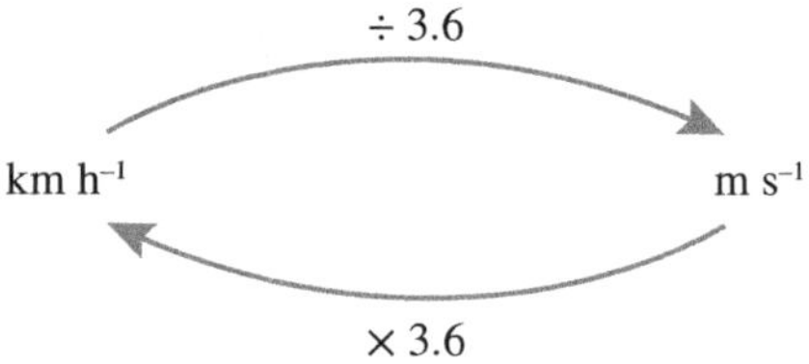

FIGURE 1.5 Unit conversion

To convert from m s^{-1} to km h^{-1}, multiply by 3.6.

To convert from km h^{-1} to m s^{-1}, divide by 3.6.

ACCELERATION

Motion that involves speeding up, slowing down, or changing direction involves acceleration. **Acceleration** is a measure of how quickly velocity changes. It can be calculated if you know the initial velocity and final velocity, and the time taken for the change in velocity.

The formulae for change in speed and change in velocity are as follows:

The change in speed is a scalar calculation:

$$\Delta v = \text{final speed} - \text{initial speed} = v - u$$

The change in velocity is a vector calculation:

$$\Delta \vec{v} = \text{final velocity} - \text{initial velocity} = \vec{v} - \vec{u}$$

If the final velocity is smaller than the initial velocity, the change in velocity is negative, which indicates that the object is slowing down in the direction of travel. This is known as deceleration.

Acceleration is a vector. The average acceleration of an object, $\vec{a}_{av}$, is defined as the rate of change of velocity:

$$\vec{a}_{av} = \frac{\text{change in velocity}}{\text{time taken}} = \frac{\Delta \vec{v}}{\Delta t} = \frac{\vec{v} - \vec{u}}{\Delta t}$$

ISBN 978 1 4886 1935 9

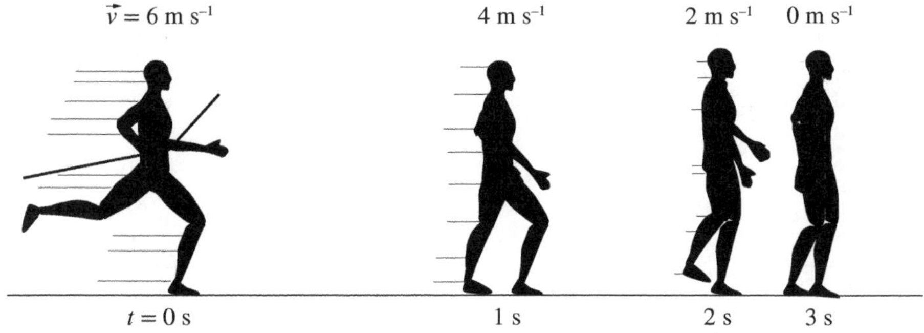

FIGURE 1.6 Negative acceleration, also known as deceleration

The standard unit for acceleration is metres per second squared (m s^{-2}).

In Figure 1.6, the runner decreased their velocity by $4 - 6 = -2\,\text{m s}^{-1}$ in the first second of motion after crossing the finish line. In the next second the runner again slowed down by $-2\,\text{m s}^{-1}$, and again in the third second. This means that during each second their velocity decreased by $2\,\text{m s}^{-1}$, so their acceleration was $-2\,\text{m s}^{-1}$ per second, or $-2\,\text{m s}^{-2}$.

GRAPHING POSITION, VELOCITY AND ACCELERATION OVER TIME

Graphs and tables are often used to represent the motion of an object. Position, velocity or acceleration are plotted on the *y*-axis, against time on the *x*-axis.

A position–time graph shows the position of an object at any given time. Other information about the object's motion can also be derived from a position–time graph:

- the displacement, which is the difference in *y* values between two points on the graph
- the average velocity, which is the gradient of the graph between two points
- the instantaneous velocity, which is the gradient at one point in time (or the tangent to the curve if the graph is not straight).

A velocity–time graph shows the velocity of an object at any given time. Other information can be also be derived from a velocity–time graph:

- the displacement, which is the total area between the graph and the *x*-axis, measured between two points on the graph
- the average acceleration, which is the gradient between two points on the graph
- the instantaneous acceleration, which is the gradient of the graph at one point in time (or the tangent to the curve if the graph is not straight).

The area under an acceleration–time graph is the change in velocity of the object. This is calculated using the same method as determining the displacement from a velocity–time graph.

Consider the position–time graph shown in Figure 1.7, for an object moving in a straight line. The origin of the graph represents the object's initial position.

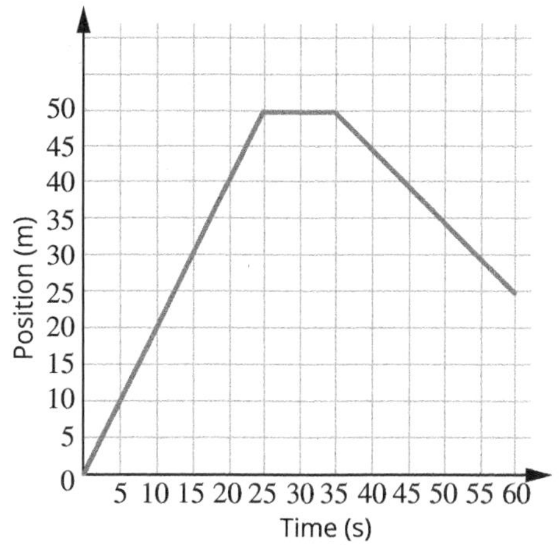

FIGURE 1.7 An example of a position–time graph

The displacement can be determined by reading points off the graph.

$$\begin{aligned}\vec{s} &= \text{final position} - \text{initial position}\\ &= +25 - 0\\ &= +25\,\text{m}\end{aligned}$$

The average velocity can be determined from the gradient, which is the change in displacement over time. For example, in the first 25 seconds the object travelled 50 m in the positive direction, so its average velocity during this time was $\frac{+50\,\text{m}}{25\,\text{s}} = +2\,\text{m s}^{-1}$.

Now consider Figure 1.8, which shows a velocity–time graph for an object moving in a straight line.

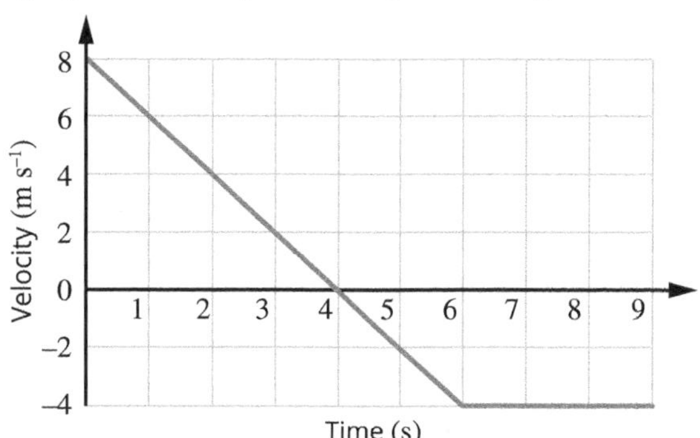

FIGURE 1.8 An example of a velocity–time graph

The average acceleration of the object can be determined from the gradient of the slope, which is the change in velocity over time. For example, in Figure 1.8 during the first 6 seconds of motion the velocity changed from +8 m s^{-1} to −4 m s^{-1}, so:

$$\begin{aligned}\vec{a}_{av} &= \frac{\vec{v}-\vec{u}}{\Delta t}\\ &= \frac{-4-8}{6}\\ &= -2\,\text{m}\,\text{s}^{-2}\end{aligned}$$

To calculate the displacement during the whole 9 seconds, the area can be split into two sections: the area above the x-axis (positive) and the area below the x-axis (negative). The area above the x-axis is a triangle, and the area below the x-axis is a trapezium consisting of a triangle (T) and a rectangle (R).

$$\begin{aligned}\text{area}_{\text{above}} &= \frac{1}{2}(b\times h)\\ &= \frac{1}{2}(4\times +8)\\ &= +16\,\text{m}\end{aligned}$$

$$\begin{aligned}\text{area}_{\text{below}} &= \frac{1}{2}(b_T\times h_T)+(b_R\times h_R)\\ &= \frac{1}{2}(2\times -4)+(3\times -4)\\ &= -4+(-12)\\ &= -16\,\text{m}\end{aligned}$$

The total displacement is +16 − 16 = 0 m. (But note that the distance travelled is 16 + 16 = 32 m.)

EQUATIONS OF MOTION

In situations involving straight-line (rectilinear) motion, the equations of motion can be used to find an unknown variable using known variables. These equations can only be used in situations where there is constant acceleration.

 The equations of motion:

- $\vec{v} = \vec{u} + \vec{a}t$
- $\vec{s} = \vec{u}t + \frac{1}{2}\vec{a}t^2$
- $\vec{s} = \vec{v}t - \frac{1}{2}\vec{a}t^2$
- $\vec{v}^2 = \vec{u}^2 + 2\vec{a}\vec{s}$
- $\vec{s} = \frac{1}{2}(\vec{u} + \vec{v})t$

where:

$\vec{s}$ is the displacement (in m)

$\vec{u}$ is the initial velocity (in m s^{-1})

$\vec{v}$ is the final velocity (in m s^{-1})

$\vec{a}$ is the acceleration (in m s^{-2})

t is the time (in s).

In solving for the unknown variable, it is important to identify all the known variables in order to select the appropriate equation. For example if the unknown variable is acceleration and the initial velocity, final velocity and displacement are known, the equation $\vec{v}^2 = \vec{u}^2 + 2\vec{a}\vec{s}$ can be used to find the acceleration.

A sign and direction convention for the motion needs to be used with these equations. For example, if the motion is along a north–south line, north is usually taken to be the positive direction.

VERTICAL MOTION

If you throw an object upwards, it will eventually fall back to the ground because of the Earth's gravity. If air resistance is ignored, all objects falling freely near the Earth will move with the same constant acceleration. The acceleration due to gravity is represented by g and is equal to 9.8 m s^{-2} downwards (i.e. towards the centre of the Earth). The motion of an object thrown or fired into the air is often called projectile motion.

The equations of motion can be used to solve vertical motion problems. Upwards is usually considered to be the positive direction. For example, consider an object being thrown vertically upwards with an initial velocity of 30 m s^{-1}. In this problem the only variable given is the initial velocity, but there is also a downwards acceleration of 9.8 m s^{-2} due to gravity. Near the Earth's surface, objects decelerate at 9.8 m s^{-2} going up and accelerate at this same rate going down. There are three other pieces of information that can be useful in solving problems about projectile motion:

- time taken to go up = time taken to come down
- final velocity = – initial velocity
- at the highest point, $\vec{v}$ = 0 m s^{-1}.

Motion on a plane

Analysing motion in one dimension requires only simple mathematics. However, vectors in two dimensions may involve a change in direction, such as walking up a hill or running around a block. In theses cases, more complex methods are needed to describe and analyse motion.

VECTORS IN TWO DIMENSIONS

Graphical analysis

The head-to-tail method used for adding and subtracting vectors in one dimension can also be used to add and subtract vectors in two dimensions. The difference is that the direction of the vectors must be indicated by a bearing or angle instead of a plus or minus sign. Compass bearings (such as south-west or S45°W), true bearings (such as 225°T) or simple angles are used in problems to indicate directions.

In the example in Figure 1.9, the vectors are drawn to a scale of 1 cm for every 5 N. A protractor is needed to draw the 45° angle and measure the angle of the resultant. This method can be used to add or subtract any number of vectors.

 ISBN 978 1 4886 1935 9

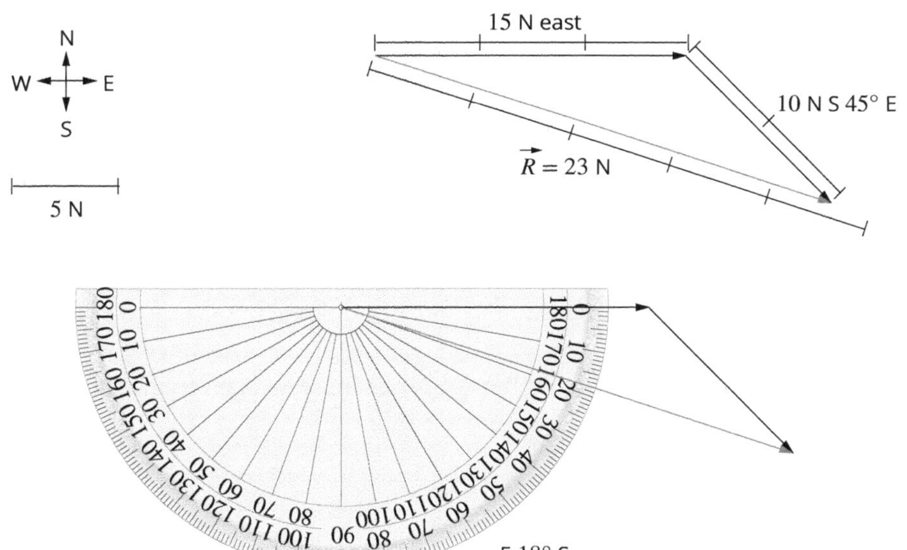

FIGURE 1.9 Graphical method for adding vectors in two dimensions

When only two vectors are involved, an alternative graphical method is to construct a parallelogram from the vectors. In this method the vectors are drawn tail-to-tail. The resultant is then a line drawn from the tails of the vectors to the opposite vertex of the parallelogram. An example of this method is shown in Figure 1.10, with the resultant vector represented by $\vec{R}$.

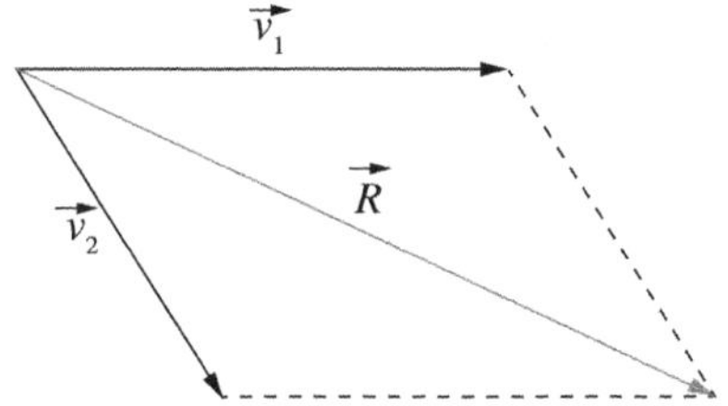

FIGURE 1.10 The parallelogram method for finding a resultant vector

Algebraic analysis

Vectors at right angles to each other can be added and subtracted using Pythagoras' theorem and the trigonometric ratios of a right-angled triangle, as shown in Figure 1.11.

To calculate the magnitude of the resultant vector, Pythagoras' theorem is used:

$$a^2 + b^2 = c^2$$

where c is the length of the hypotenuse, and a and b are the lengths of the other two sides.

To find the direction of the resultant vector, the appropriate trigonometric ratio is used:

$$\sin\theta = \frac{\text{opposite}}{\text{hypotenuse}}$$

$$\cos\theta = \frac{\text{adjacent}}{\text{hypotenuse}}$$

$$\tan\theta = \frac{\text{opposite}}{\text{adjacent}}$$

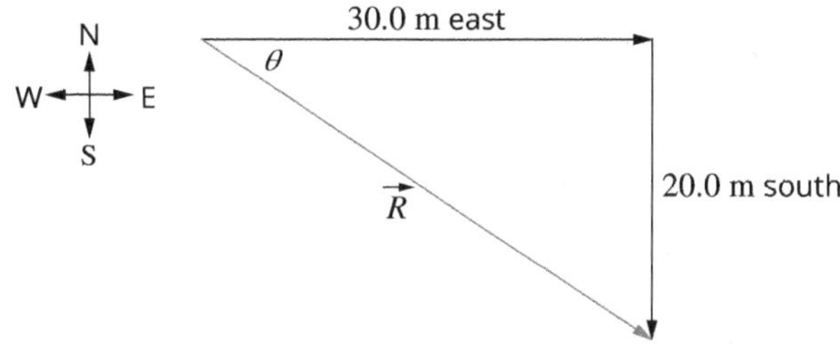

FIGURE 1.11 Geometric method for adding vectors at right angles to each other

VECTOR COMPONENTS

Consider someone pulling a cart, as shown in Figure 1.12. The force exerted on the cart can be resolved (broken up) into two perpendicular components: a vertical component and a horizontal component.

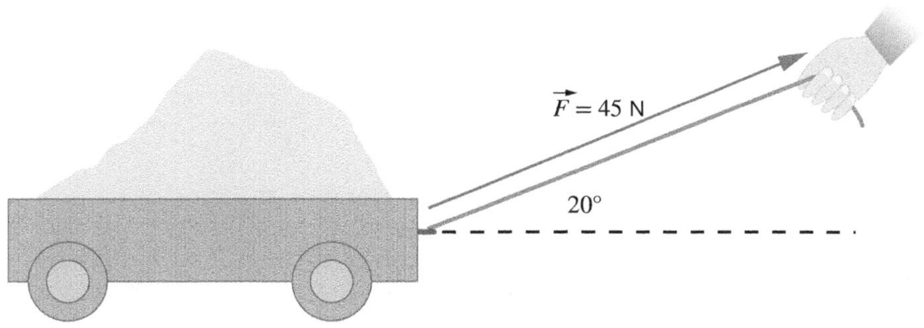

FIGURE 1.12 Vector components: pulling a cart

The perpendicular components are at right angles to each other and therefore make a right-angled triangle, with the original force vector forming the hypotenuse. This means that some properties of right-angled triangles can be used to find the magnitudes of the components:

- Any component vector must be smaller in magnitude than the original vector, because the hypotenuse is the longest side of the triangle.
- Graphically, a right-angled triangle vector diagram can be drawn with the original vector as the hypotenuse and the perpendicular components drawn from the tail to the head of the original vector.
- Algebraically, the magnitudes of the perpendicular components can be found using the trigonometric ratios.

Resolving vectors into components is an important tool for analysing the effects of two or more forces acting on an object. For example, if an additional friction force is applied to the wheels in Figure 1.12, it affects only the horizontal component. The new horizontal component can be easily calculated and then recombined with the unchanged vertical component to find the new resultant force vector.

RELATIVE MOTION

If a car heading north passes a pedestrian waiting to cross the road, the pedestrian would believe that they are stationary and the car is moving north. But the driver of the car could say that the car is stationary and the pedestrian is moving south. In physics, both of these views are correct.

How motion is described depends on the **frame of reference** that is used. An object's frame of reference is the coordinate system that is used to describe the motion of an object. It can be stationary (like the pedestrian's frame of reference) or it can be moving (like the driver's frame of reference).

When describing the velocity of one object (A) relative to another object (B), the vector notation used is $\vec{v}_{AB}$. The first subscript refers to the object that is considered to be stationary, and the second subscript refers to the object considered to be in motion. It is useful to note that $\vec{v}_{AB} = -\vec{v}_{BA}$.

To calculate the relative velocity of two objects, a frame of reference is needed. In most problems this is defined by a surface (such as the ground or a road) or a point (such as a stationary object or observer).

The velocity of A relative to B is then given by:

$$\vec{v}_{AB} = \vec{v}_{AC} + \vec{v}_{CB}$$

In words this is:

The velocity of A relative to B is the sum of the velocity of A relative to C and the velocity of C relative to B.

Notice that the subscripts on the right-hand side are ordered so that if you crossed out each C you would be left with AB.

Consider two cars travelling in the same direction on a freeway. Car A is travelling at 90 km h^{-1} relative to the road, and car B is travelling at 100 km h^{-1} relative to the road, what is the velocity of car A relative to car B?

The first step is to identify what information has been given. Let R = road, A = car A and B = car B.

$\vec{v}_{AR} = 90\,\text{km h}^{-1}$, $\vec{v}_{BR} = 100\,\text{km h}^{-1}$, $\vec{v}_{AB} = ?$

$$\vec{v}_{AB} = \vec{v}_{AR} + \vec{v}_{RB}$$

Because $\vec{v}_{BR}$ is given instead of $\vec{v}_{RB}$, use $\vec{v}_{RB} = -\vec{v}_{BR}$.

So:

$$\begin{aligned}\vec{v}_{AB} &= \vec{v}_{AR} + (-\vec{v}_{BR})\\ &= 90 - 100\\ &= -10\,\text{km h}^{-1}\end{aligned}$$

Therefore car A is travelling at -10 km h^{-1} relative to car B.

In this example the vectors are in one dimension. When considering relative velocity in two dimensions, the same process is followed but Pythagoras' theorem and trigonometric ratios may be needed.

Consider a boat (B) moving forward at 4 m s^{-1} across a river (R) where the current is 6 m s^{-1}, as shown in Figure 1.13. What is the velocity of the boat relative to the ground at the river's edge (G)?

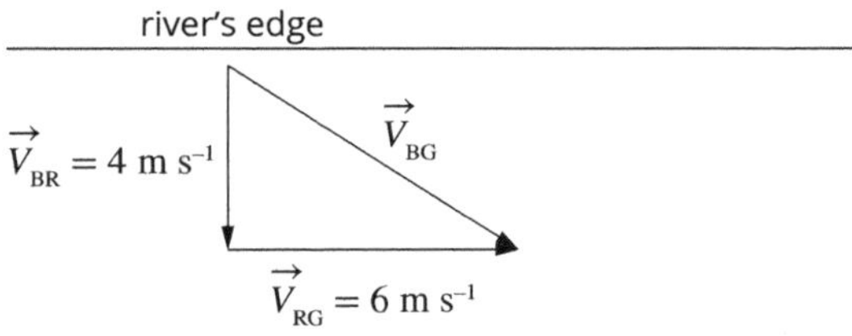

FIGURE 1.13 Analysing relative motion in two dimensions

$\vec{v}_{BR} = 4\,\text{m s}^{-1}$, $\vec{v}_{RG} = 6\,\text{m h}^{-1}$, $\vec{v}_{BG} = ?$

$$\vec{v}_{BG} = \vec{v}_{BR} + \vec{v}_{RG}$$

Using Pythagoras' theorem and trigonometry:

$$\begin{aligned}\vec{v}_{BG} &= \sqrt{6^2 + 4^2}\\ &= 7.2\,\text{m s}^{-1}\\ \theta &= \tan^{-1}\frac{4}{6}\\ &= 33.7^\circ\end{aligned}$$

$\therefore \vec{v}_{BG} = 7.2$ m s^{-1} at an angle of 33.7° to the ground at the river's edge.

ISBN 978 1 4886 1935 9

WORKSHEET 1.1

Knowledge review—motion

1 A girl on a skateboard travels 16 m down the street in 2 s. What is her speed? In general, speed is found by dividing distance by the time taken.

2 The following diagram shows the journey of an ant from start to end.

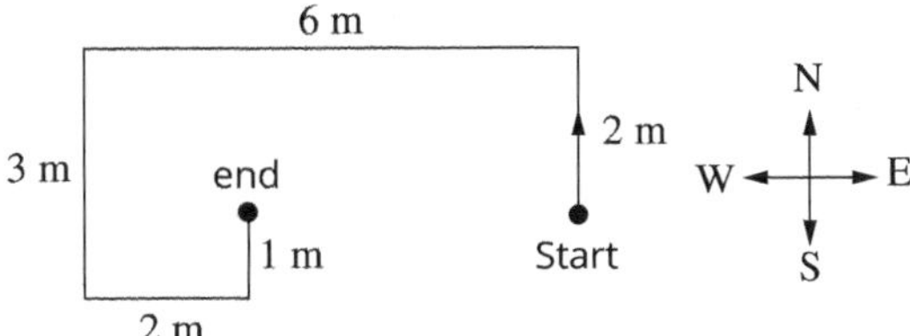

a What distance did the ant travel?

b If the journey took 56 seconds altogether, what was the ant's average speed?

c What is its total displacement from the start to the end? (Remember that displacement is the difference between the final position and the starting position, and it is a vector quantity.)

3 A truck is travelling at 100 km h^{-1} from Wangaratta to Wagga Wagga, a distance of 200 km. How long will it take:

a in hours?

b in minutes?

4 Using your usual trip to school, explain the difference between average speed and instantaneous speed.

5 Examine the three ticker tapes below. The start label shows where the first dot was made.

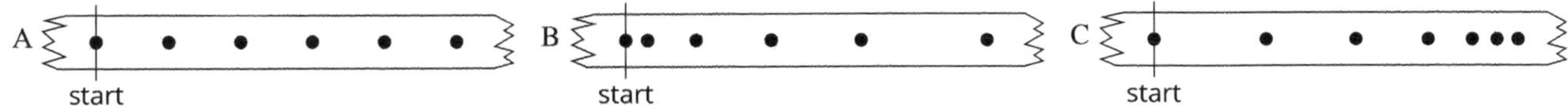

a When these tapes were made, did the tape move to the left or the right?

b Which tape shows acceleration?

c Which one shows deceleration?

d Which one shows constant speed?

WORKSHEET 1.2

Vectors in the classroom

Any motion between two points can be described in terms of scalar and vector quantities. Describing a journey purely in terms of scalar quantities makes it very difficult, if not impossible, to determine the final destination. By adding direction and knowing the starting point, the final destination can be found.

Below is a 1 cm grid representing your classroom. The front and back of the classroom are shown and a 'classroom compass', using directions of forwards/backwards and left/right, with which to indicate direction. Your task is to plot a path on the grid that would allow you to navigate from your current seat to the teacher's desk, wherever that might be in the room. There are some rules that apply:

- Determine a scale that would work for your classroom. 1 cm = ____ metres
- Use only straight lines. Show these as displacement vectors, labelled with the distance in metres and the direction based on the classroom compass directions.
- You may only travel left, right, forwards or backwards: no diagonals!
- Add your displacement vectors head to tail.
- Estimate your distance travelled; your teacher may allow you to measure directly but ask before moving around. The nearest half metre or metre will do.
- You must avoid any desks, students and other obstacles in the room (you might like to draw these in to assist your plot). Plot a path that you could really walk.
- Your path must use a minimum of five displacement vectors, and a maximum of ten.

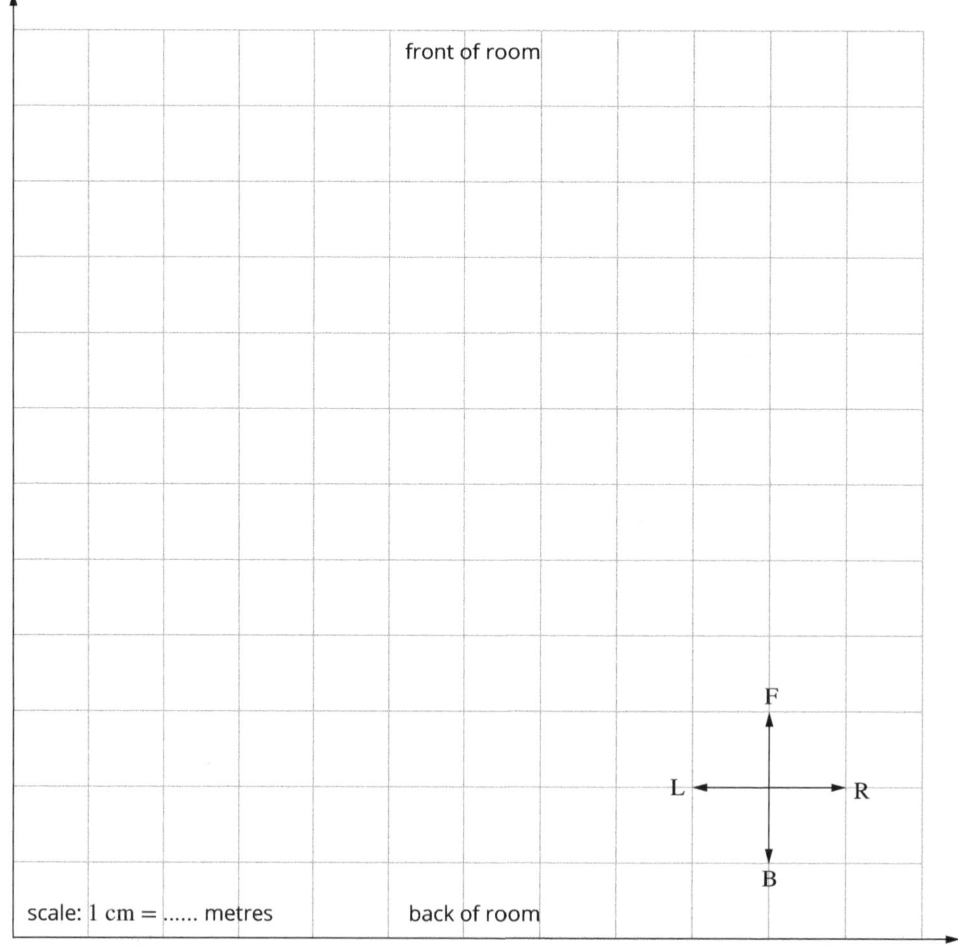

ISBN 978 1 4886 1935 9

WORKSHEET 1.2

1 Tabulate your displacement vectors in the table below. Treat the directions of forwards and right as positive, and left and backwards as negative.

Step	Displacement	Vertical component	Horizontal component
(Example)	3 m, left	0	–3
1			
2			
3			
4			
5			
6			
7			
8			
9			
10			
Total			

2 Calculate the total resultant vertical and horizontal components.

3 Draw the displacement vector from your position to that of your teacher on the classroom grid. Use a protractor and ruler to determine the distance and direction of this vector and label the vector.

Extension

4 Calculate the total displacement (resultant vector) from your desk to your teacher's desk.

5 Compare your calculated resultant vector with that measured from your diagram.

RATING MY LEARNING	My understanding improved	Not confident ◄——► Very confident ○ ○ ○ ○ ○	I answered questions without help	Not confident ◄——► Very confident ○ ○ ○ ○ ○	I corrected my errors without help	Not confident ◄——► Very confident ○ ○ ○ ○ ○

WORKSHEET 1.3

Speed versus velocity

In everyday language, speed and velocity are often used interchangeably. In physics there is an important distinction between the two terms.

1 Speed is a ________________ quantity. It is completely described by its ________________.

Velocity is a ________________ quantity. It is completely described by its ________________ and ________________.

2 A GPS offers a teacher three potential paths to travel by car between Broken Hill High School and a local take-away restaurant for lunch. To arrive at the same time, on which route (A, B or C in the diagram) would the:

a average speed be higher?

____

b average velocity be higher?

____

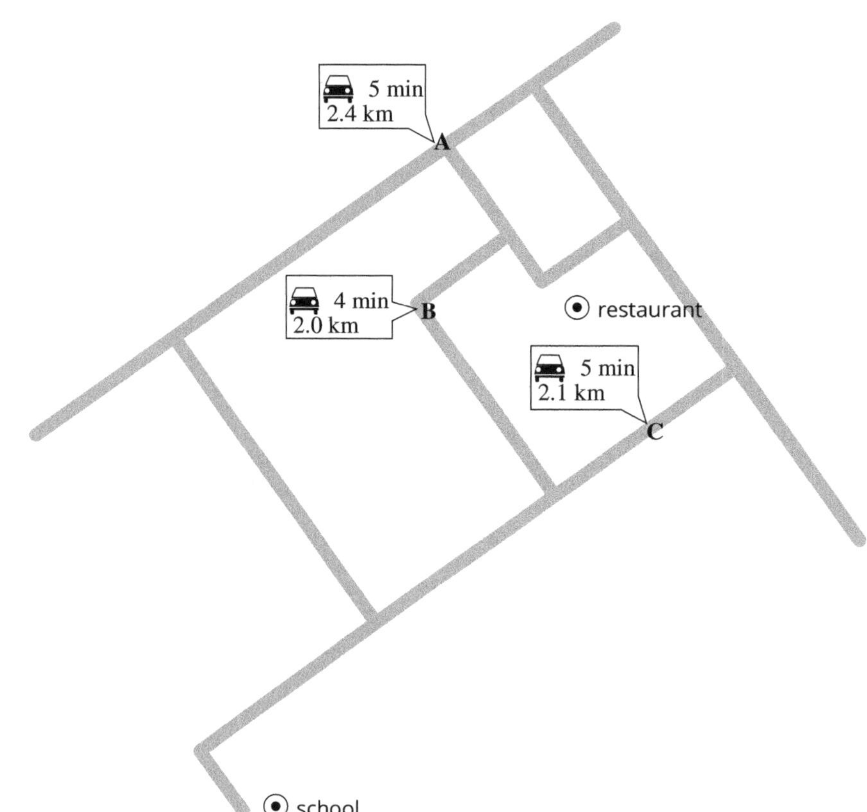

3 After buying lunch, the teacher in Question 2 returns along the route that has the highest average speed, taking the same time. Based on the full trip to and from the restaurant, what would be the teacher's:

a average speed in $m\,s^{-1}$ and $km\,h^{-1}$?

____

____

b average velocity in $m\,s^{-1}$ and $km\,h^{-1}$?

____

____

(Show your working and/or reasoning in each part.)

 ISBN 978 1 4886 1935 9

4 An athlete is completing a set of high-intensity shuttle runs of increasing distance, as shown in the figure below.

The lines are 10 metres apart.

She completes each run from the origin in the time shown in the table.

The total test is completed when each run, from O to A and back, O to B and back, and O to C and back, is completed.

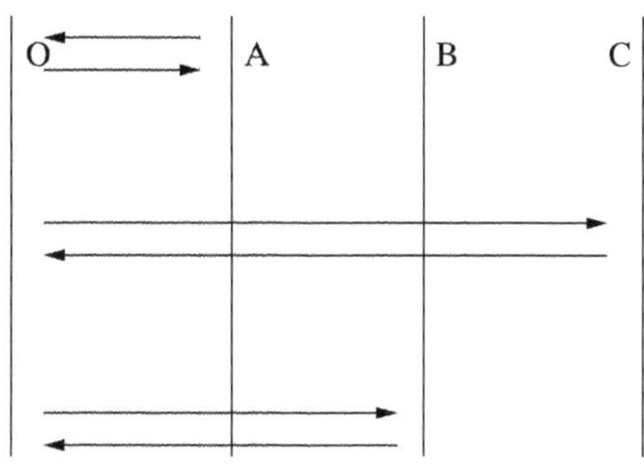

Complete the table below for the missing quantities.

Step	Distance (m)	Time (s)	Average speed ($m\,s^{-1}$)	Average velocity ($m\,s^{-1}$)
O–A	10	1.6		
O–A–O		3.5		
O–B		3.4		
O–B–O		7.2		
O–C		5.2		
O–C–O		10.7		
Total Time				

5 On the grid below, draw a distance–time graph and a displacement-time graph showing the athlete's total test.

RATING MY LEARNING	My understanding improved	Not confident ◄——► Very confident ○ ○ ○ ○ ○	I answered questions without help	Not confident ◄——► Very confident ○ ○ ○ ○ ○	I corrected my errors without help	Not confident ◄——► Very confident ○ ○ ○ ○ ○

WORKSHEET 1.4

Speed–time versus velocity–time

The following figure shows an incomplete graph of an object's motion.

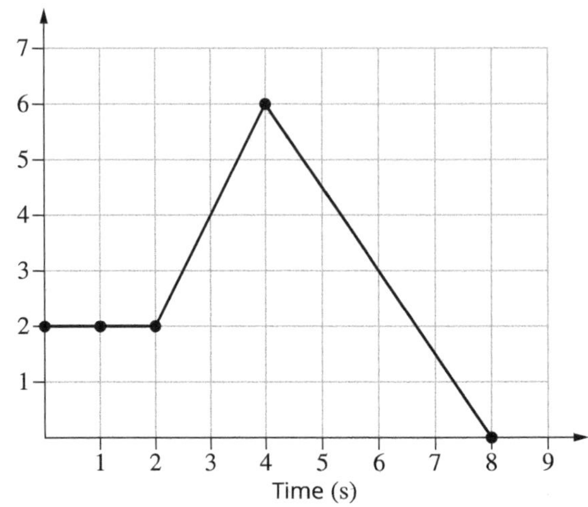

The unit for the graph's vertical axis is $m\,s^{-1}$.

The positive gradient between 2 and 4 seconds represents the object's acceleration.

1 What type of graph is this?

A a velocity–time graph

B a speed–time graph

C either a speed–time or a velocity–time graph

2 Justify your response to Question 1.

3 What is the magnitude of the largest acceleration?

4 At what time, or times, is the object stationary? Justify your answer.

5 At what time, or times, is the object moving backward? Justify your answer.

ISBN 978 1 4886 1935 9

WORKSHEET 1.4

The following figure is a velocity–time graph for an object moving in a straight line.

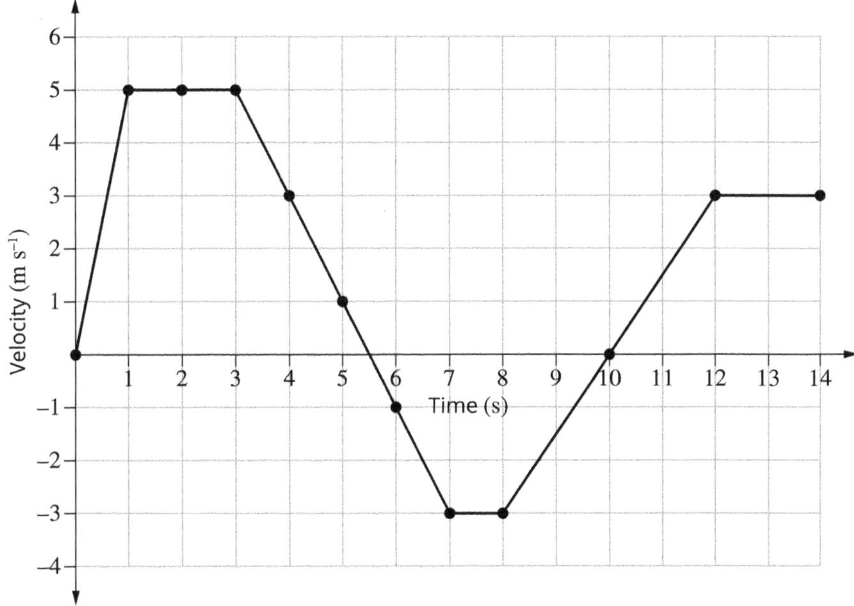

6 At what time, or times, was the object stationary?

7 What was the velocity of the object at $t = 7\,s$?

8 What was the acceleration of the object at $t = 5\,s$?

9 When was the first time that the object was moving backwards?

10 Describe the motion of the object between:

a $t = 8\,s$ and $t = 10\,s$

b $t = 10\,s$ and $t = 12\,s$

11 What was the total displacement of the object from $t = 0\,s$ to $t = 14\,s$?

RATING MY LEARNING	My understanding improved	Not confident ◄—► Very confident ○ ○ ○ ○ ○	I answered questions without help	Not confident ◄—► Very confident ○ ○ ○ ○ ○	I corrected my errors without help	Not confident ◄—► Very confident ○ ○ ○ ○ ○

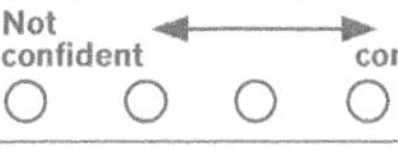

WORKSHEET 1.5

Position–time and velocity–time graphs

The left-hand side of the page shows a series of position–time graphs. On the right-hand side is a series of velocity–time graphs.

1 For each position–time graph, write the number of the corresponding velocity–time graph.

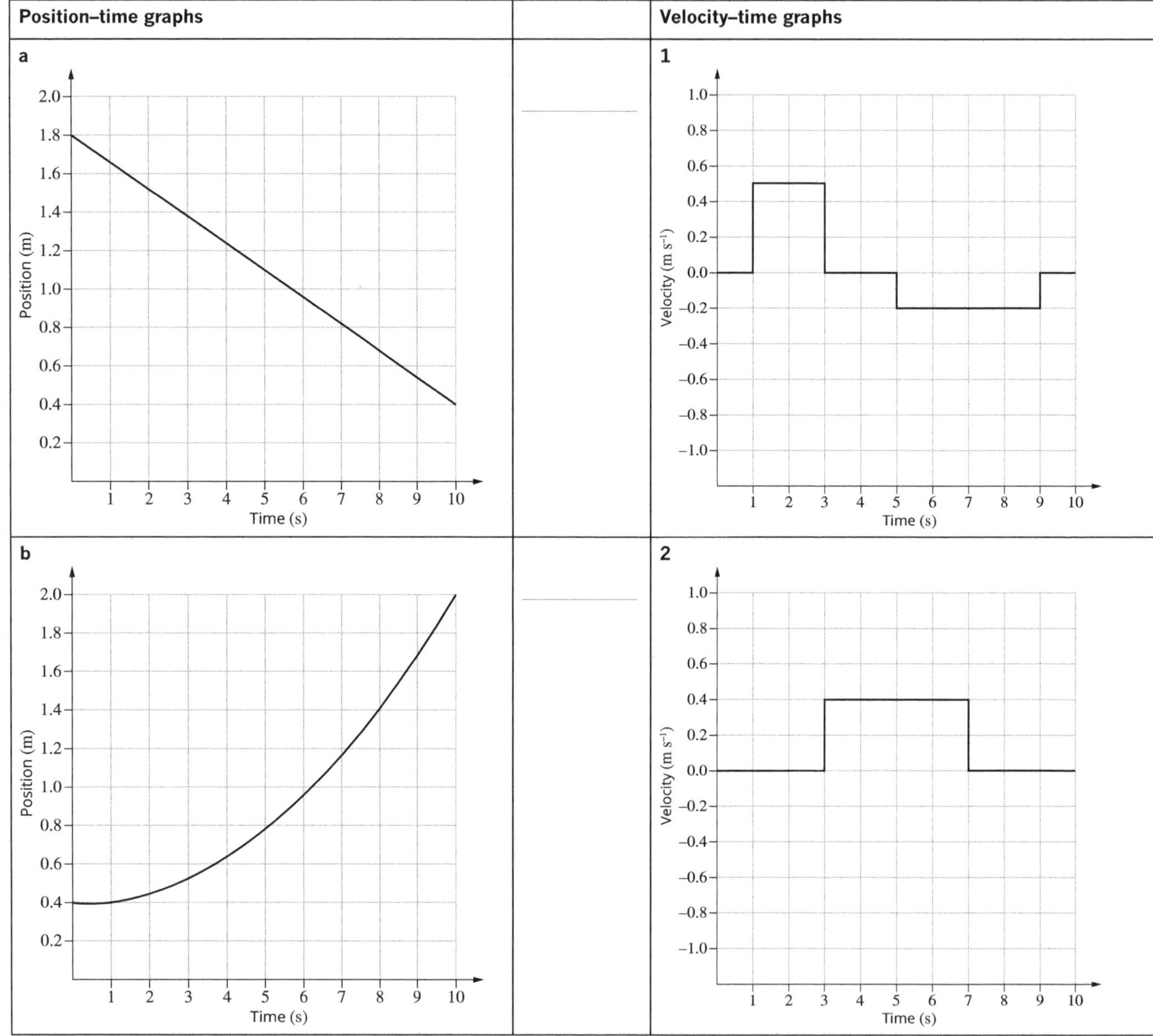

ISBN 978 1 4886 1935 9

c Position (m) 0.2–2.0 vs Time (s) 1–10	________	**3** Velocity (m s⁻¹) −1.0–1.0 vs Time (s) 1–10
d Position (m) 0.2–2.0 vs Time (s) 1–10	________	**4** Velocity (m s⁻¹) −1.0–1.0 vs Time (s) 1–10
e Position (m) 0.2–2.0 vs Time (s) 1–10	________	**5** Velocity (m s⁻¹) −1.0–1.0 vs Time (s) 1–10

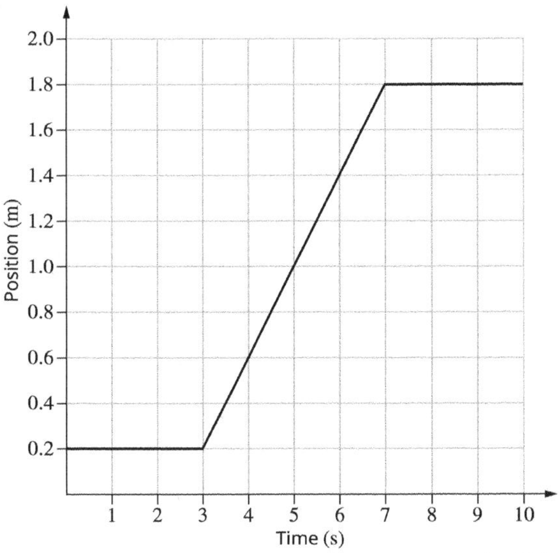

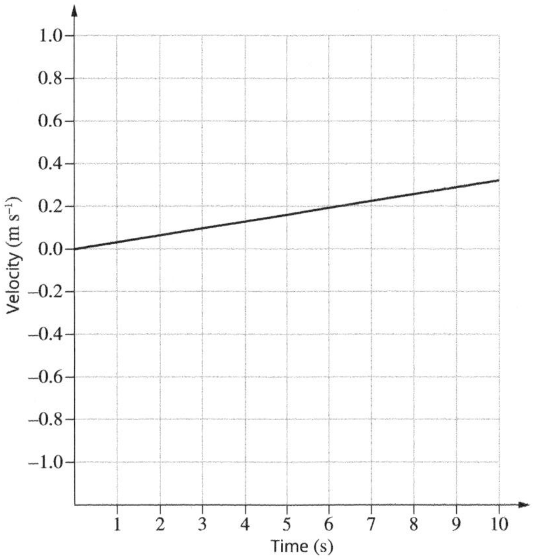

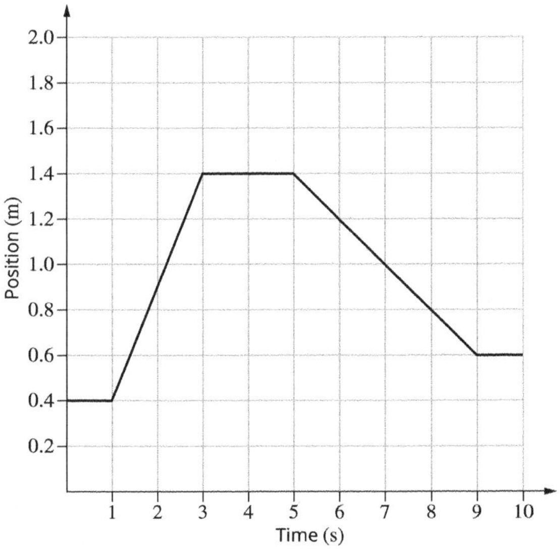

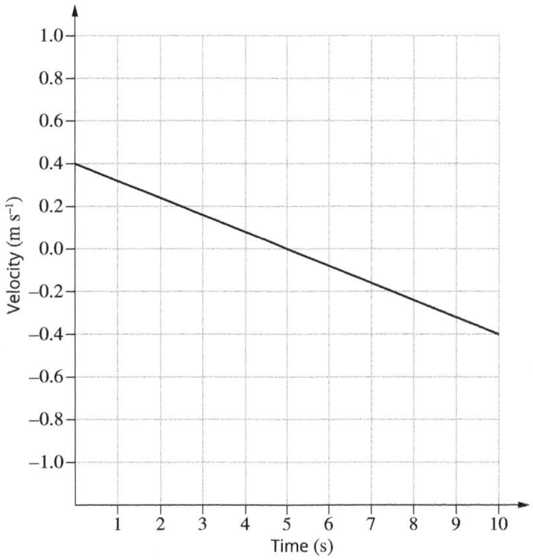

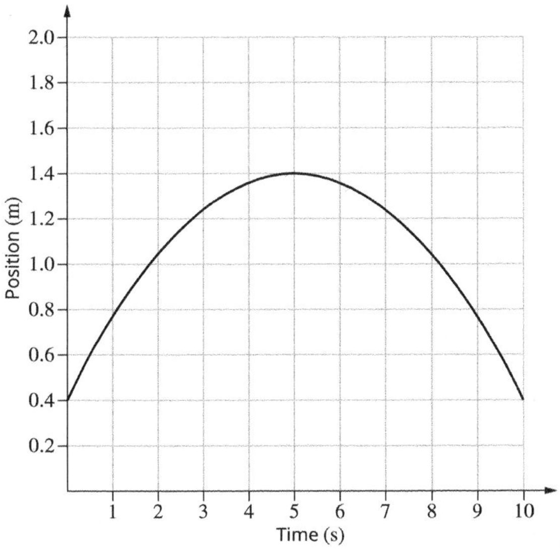

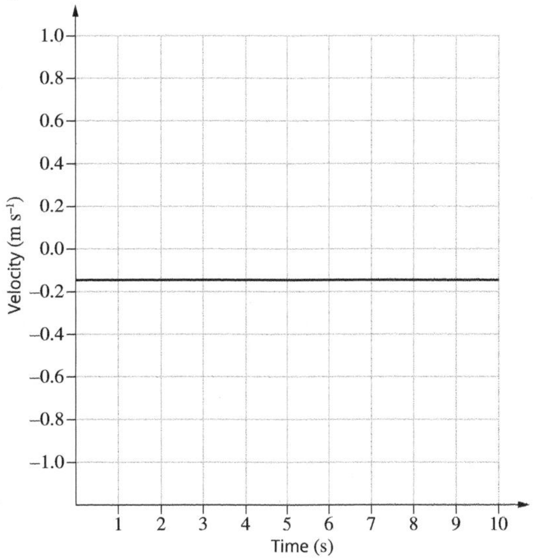

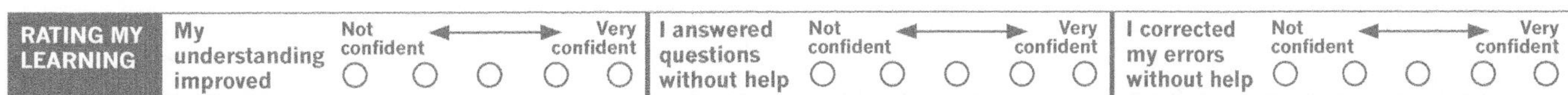

ISBN 978 1 4886 1935 9

WORKSHEET 1.6

Free falling—a special case of straight-line motion

The equations of motion give the relationships between time, distance, displacement, speed, velocity and acceleration in rectilinear (straight line) motion. The three standard forms are as follows:

$$\vec{s} = \vec{u}t + \frac{1}{2}\vec{a}t^2$$

$$\vec{v} = \vec{u} + \vec{a}t$$

$$\vec{v}^2 = \vec{u}^2 + 2\vec{a}\vec{s}$$

1 Explain the significance of the arrow above some of the quantities.

2 Name each of the vector quantities in the equations, their SI unit of measurement, and the corresponding scalar quantity.

$\vec{s}$

$\vec{u}$

$\vec{v}$

$\vec{a}$

3 The equations of motion for rectilinear motion can be applied only if a certain condition is met. What is this condition?

4 A steel sphere is projected vertically upwards with an initial velocity of 10 m s^{-1} at the locations shown in the following table. Calculate the maximum height the ball will reach, assuming it is projected from an initial height of 1.00 m above the surface.

Location	$\vec{g}$ (m s^{-2})	Final height (m)
Equator	9.780	
South Pole	9.832	
Sydney	9.797	
Moon	1.625	
Mars	3.711	
Jupiter	24.5	
Venus	8.87	

5 A robot on Titan, one of Jupiter's moons, drops a steel sphere from 100 metres above the surface in order to confirm the value of the acceleration due to gravity $\vec{g}$. The ball takes 12.16 s to reach the surface. Assuming the initial velocity is zero and there is no air resistance, what is the value of $\vec{g}$ on Titan? Show your working.

RATING MY LEARNING	My understanding improved	Not confident ◄——► Very confident ○ ○ ○ ○ ○	I answered questions without help	Not confident ◄——► Very confident ○ ○ ○ ○ ○	I corrected my errors without help	Not confident ◄——► Very confident ○ ○ ○ ○ ○

 ISBN 978 1 4886 1935 9

WORKSHEET 1.7

Race analysis—a study of relative motion

Australia was represented by two competitors for the 200 m women's freestyle final in the 2016 Olympic Games. Emma McKeon led the field for the first two laps, turning under world record pace at the 100 m mark. She faded on the third 50 m lap and was finally placed 3rd for the bronze medal. Bronte Barratt was placed a very creditable 5th, finishing only 0.33 s outside the medal positions.

Their split times are shown in the following table.

Competitor	50 m split (s)	100 m split (s)	150 m split (s)	200 m split (s)
Emma McKeon	26.64	55.37	85.37	114.92
Bronte Barratt	27.35	56.40	86.02	115.25

1 Calculate the average speed for each lap for each competitor. Record your answers in the table below.

Competitor	Lap 1 ($m\,s^{-1}$)	Lap 2 ($m\,s^{-1}$)	Lap 3 ($m\,s^{-1}$)	Lap 4 ($m\,s^{-1}$)
Emma McKeon				
Bronte Barratt				

2 Using a vector diagram and relevant calculations, find Emma's velocity compared to Bronte's during Lap 1. Adopt a suitable direction as positive to show your answer.

3 Using a vector diagram, and appropriate calculations, find the velocity of Bronte relative to Emma for laps 3 and 4.

4 Based on the average speed for each lap, what was the greatest difference in the magnitude of the relative velocity between Emma and Bronte during the race? Use a vector diagram and appropriate calculations to show your working.

RATING MY LEARNING	My understanding improved	Not confident ◄——► Very confident ○ ○ ○ ○ ○	I answered questions without help	Not confident ◄——► Very confident ○ ○ ○ ○ ○	I corrected my errors without help	Not confident ◄——► Very confident ○ ○ ○ ○ ○

WORKSHEET 1.8

Analysing motion in two dimensions

A soccer team uses GPS trackers as part of their pre-season player assessment. Each player wears a special vest fitted with a tracker that records time, position, speed and direction during game play, as shown in the figure below. Analysis of the data provides key metrics for player performance and positioning during play.

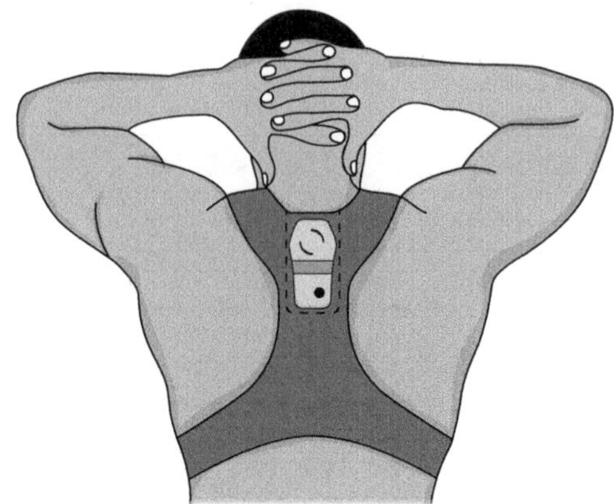

The data for one of the players during a short period of play is shown in the table below.

- The player has taken the kick-off at the start of play and then moved off around the playing field.
- The table shows the player's speed, full circle bearing and the time travelled for each individual move (not the total running time).
- A new entry is made for each change of direction. Each entry can be assumed to be in a straight line.

Position	Time (s)	Speed, v ($m\,s^{-1}$)	Full circle bearing (°T)	Distance travelled (m)	Angle (°) from north or south	North–south component (m)	East–west component (m)
1	1.2	5.0	270				
2	0.8	2.4	322				
3	2.5	4.3	180				
4	1.1	3.2	95				
5	2.3	0.0	0				
6	4.6	3.4	42				
7	1.3	3.1	132				
Totals		—	—		—		

1 The diagram below shows the playing field and the player's starting position. Plot each entry at a suitable scale, and draw vectors to illustrate the player's motion.

2 Draw the resultant vector for the player's displacement from his starting position to his finishing position. Measure this displacement from the vector diagram and record it on the diagram.

 ISBN 978 1 4886 1935 9

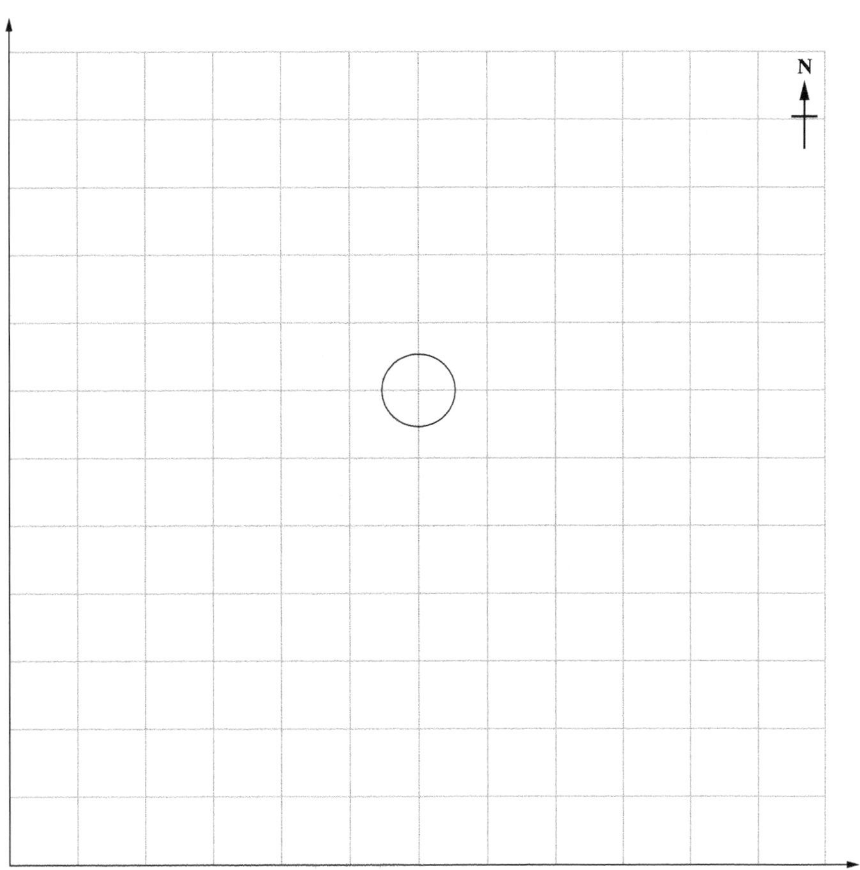

3 Using the data provided, calculate the following.

a The total distance travelled.

b The player's average speed.

c The player's final displacement from his initial position.

d The player's average velocity.

4 Compare the calculated displacement with the measured displacement.

RATING MY LEARNING	My understanding improved	Not confident  Very confident ○ ○ ○ ○ ○	I answered questions without help	Not confident  Very confident ○ ○ ○ ○ ○	I corrected my errors without help	Not confident ←→ Very confident ○ ○ ○ ○ ○

ISBN 978 1 4886 1935 9

WORKSHEET 1.9

Relative motion

The skippers of river ferries must allow for currents when navigating across a river to and from fixed points on either bank.

In the diagram below, a ferry is heading east at 8.0 m s^{-1} across a river, taking people from one side to the other. The current is flowing downstream (south) at 4.5 m s^{-1}. This situation is shown in the following diagram.

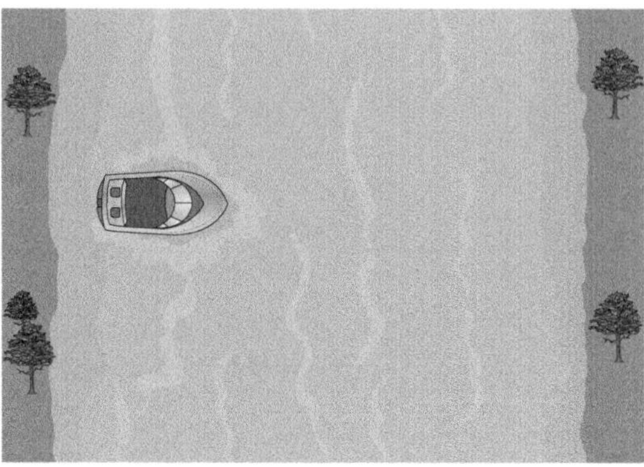

1 What is the resultant velocity of the ferry across the river? Show your working both as a vector diagram and numerically.

2 The river is 120 metres wide. How much time would the ferry take to travel from one side to the other, assuming its velocity is constant for the whole journey?

3 What distance downstream, from a position directly across from its original point of departure, would the boat reach the opposite shore?

4 If the speed of the current increased to 6.0 m s^{-1}, how much farther would the boat travel downstream?

RATING MY LEARNING	My understanding improved	Not confident ◄──► Very confident ○ ○ ○ ○ ○	I answered questions without help	Not confident ◄──► Very confident ○ ○ ○ ○ ○	I corrected my errors without help	Not confident ◄──► Very confident ○ ○ ○ ○ ○

 ISBN 978 1 4886 1935 9

WORKSHEET 1.10

Literacy review—developing a physics vocabulary

The language of science helps scientists to exchange ideas clearly across a state, a country, and even across different languages. The consistency of the language means that anyone, anywhere can convey their understanding of physics by using the same key terms. Some words might not seem to 'fit' with everyday language, but as we understand their meaning then we can convey our own understanding in the common language of science.

Many science words come from Latin or Greek. In this activity you will start a glossary of key terms for your study of physics. From your study of kinematics and using the key knowledge section of this workbook, your text and other reputable sources, complete the table below, including the origins of the words. The first two are done for you as examples.

Term	Meaning	Origin
Kinematics	The study of the motion of objects without considering the mass or the forces acting on them.	*kinematos* (Greek) – motion
Scalar	A quantity that has a magnitude but not direction.	*scalaris* (Latin) – a ladder
Vector		
Distance		
Displacement		
Speed		
Velocity		
Acceleration		
Relative		
Energy		
Components		
Plane		
Quantitative		
Qualitative		

RATING MY LEARNING	My understanding improved	Not confident ◄──► Very confident ○ ○ ○ ○ ○	I answered questions without help	Not confident ◄──► Very confident ○ ○ ○ ○ ○	I corrected my errors without help	Not confident ◄──► Very confident ○ ○ ○ ○ ○

WORKSHEET 1.11

Thinking about my learning

After completing Module 1: Kinematics, you should be able to describe, explain and apply the relevant scientific ideas. You should be able to work with data to interpret, analyse and evaluate it.

1 The table below lists the key knowledge covered in this module. Read each and reflect on how well you understand each concept. Rate your learning by shading the circle that corresponds to your level of understanding for each concept. It may be helpful to use colour as a visual representation. For example:

- green—very confident
- orange—in the middle
- red—starting to develop.

Concept focus	**Rate my learning**				
	Starting to develop ◄——► Very confident				
Difference between scalars and vectors	○	○	○	○	○
Addition and subtraction of vectors in one dimension	○	○	○	○	○
Addition and subtraction of vectors in two dimensions	○	○	○	○	○
Resolution of vectors into components	○	○	○	○	○
Analysis of graphs of displacement, velocity and acceleration over time	○	○	○	○	○
Equations of motion	○	○	○	○	○
Relative position and motion of objects in one and two dimensions	○	○	○	○	○

2 Consider where you have shaded in one of the first three circles. List any specific concepts that you found challenging in this module.

3 Write down two different strategies that you will apply to help improve your understanding of these concepts.

ISBN 978 1 4886 1935 9

PRACTICAL ACTIVITY 1.1

The ticker timer

Suggested duration: 45–90 minutes, depending on the length of tape analysed

INTRODUCTION

Mains electricity in Australia is an alternating current (AC) with a frequency of 50 Hz (50 cycles per second). This provides a very reliable means of measuring time, because the interval between each cycle is exactly 0.02 s.

PURPOSE

To investigate the relationship between position and velocity using a ticker timer.

MATERIALS

- ticker timer (AC voltage)
- power supply
- connecting wires
- ticker tape and carbon discs
- ruler

A DC-powered ticker timer with a 50 Hz hammer frequency can also be used in this activity.

PROCEDURE

1 Connect the ticker timer to the AC terminals of the power supply. Ensure that the power supply voltage is initially zero. Switch on the power supply and adjust the voltage until a regular, solid impact is recorded. Without the carbon in position, practise pulling a short length of ticker-tape (about 1 metre) through the timer at a steady speed. Practise with different speeds, and try to avoid too much variation in speed.

2 Put a carbon disc face down into the timer, with the tape underneath the carbon. Switch on the timer and pull the tape through at the same slow, steady speed. Repeat with a second piece of tape of similar length at about double the speed, then repeat again at a faster speed. Clearly label the tapes so that you can tell which trial each tape came from.

3 For each trial, discard the first part of the tape where the initial acceleration occurred. From a clear starting 'dot', count the dots on the tape and clearly mark each 5th mark. Make sure that each interval is exactly 5 dots.

4 Measure and record the total distance from the starting dot for the first 10 intervals.

DATA AND ANALYSIS

Record your data in the following table.

Interval	Total time (s)	Trial 1 position (cm)	Trial 2 position (cm)	Trial 3 position (cm)
start	0.0	0.0	0.0	0.0
1st	0.1			
2nd	0.2			
3rd	0.3			
4th	0.4			
5th	0.5			
6th	0.6			
7th	0.7			
8th	0.8			
9th	0.9			
10th	1.0			

PRACTICAL ACTIVITY 1.1

1 Use your measurements to construct a position–time graph for each trial. Because you are using an AC ticker timer and the supply has a frequency of 50 Hz, each tick represents 0.02 s. This means each interval of 5 dots represents 0.1 s.

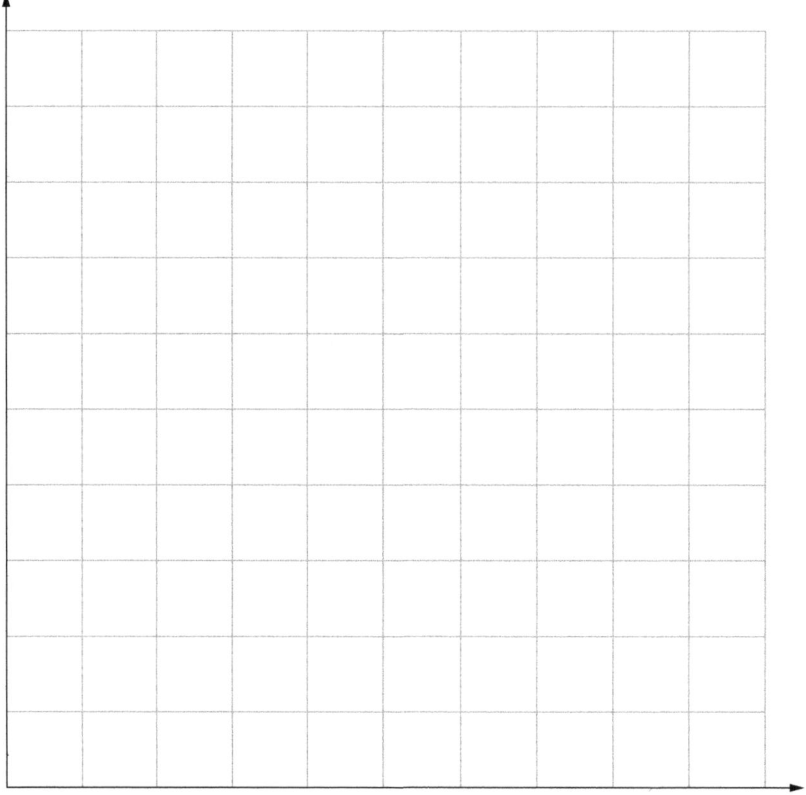

2 Using the data table or position–time graph, construct a velocity–time graph for one of the trials. Calculate the average velocity over the total distance for this trial.

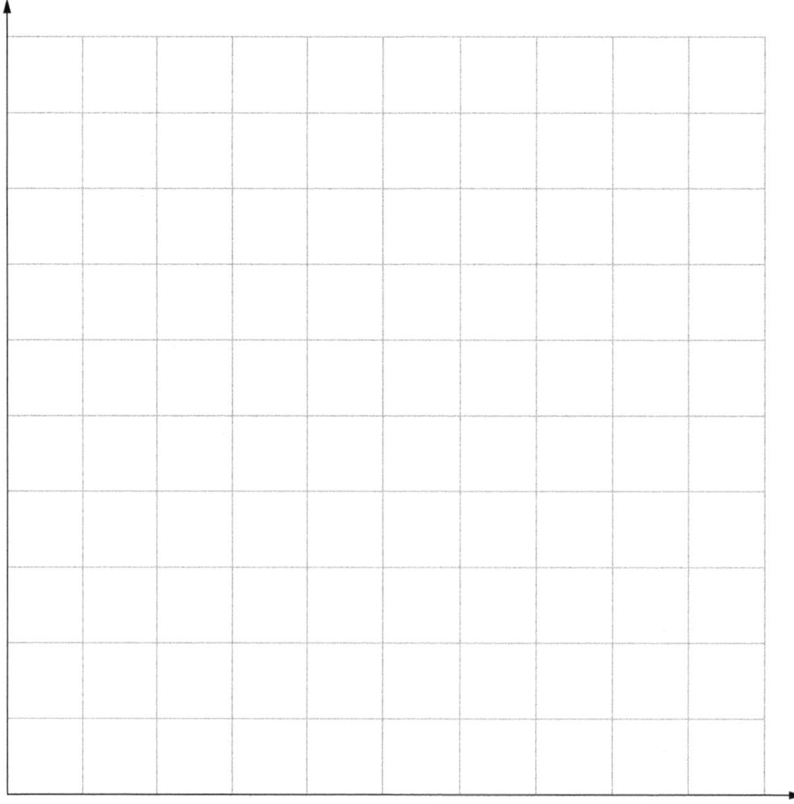

ISBN 978 1 4886 1935 9

CONCLUSION

1 Describe how you calculated velocity from the position–time graph.

2 Does the velocity–time graph show instantaneous velocity versus time or average velocity versus time? Explain.

3 Comment on the reliability and practicality of the ticker timer as a means of measuring position, time and, hence, velocity.

4 Why is it important that the hammer of the ticker timer strikes at a constant rate?

5 You were told to discard the first part of each ticker-tape because it recorded unavoidable acceleration. Why is this acceleration unavoidable?

6 What is the major source of error in this activity? What are some other possible sources of error?

RATING MY LEARNING	My understanding improved	Not confident ◄——► Very confident ○ ○ ○ ○ ○	I answered questions without help	Not confident ◄——► Very confident ○ ○ ○ ○ ○	I corrected my errors without help	Not confident ◄——► Very confident ○ ○ ○ ○ ○

ISBN 978 1 4886 1935 9

PRACTICAL ACTIVITY 1.2

The kinematics of a student

Suggested duration: 45 minutes

INTRODUCTION

The timing of intervals during races is a key indicator of performance. Graphing position against time or speed against time provides a visual indicator of performance during a race.

MATERIALS

- cones or other position markers
- 30 metre tape measure
- 6 stopwatches

PURPOSE

To investigate the motion of a student using timed intervals of measured distance.

This activity is intended as an outside activity for the whole class, but it can be conducted in a gym or hall if the weather is not suitable. It can also be conducted as a small group activity.

PROCEDURE

1 In a large, clear space, stretch out the 30 metre measuring tape on the ground in a straight line. Mark every 5 metre interval up to the 30 metre mark.

2 Station a student with a stopwatch at each 5 metre marker. Ensure each watch is zeroed and that the students are familiar with the basic operation of the stopwatch.

3 Select a student for testing and position them ready to run, walk, crawl on hand and knees, or hop from the start to the 30 metre mark. On a call of '3–2–1–GO', the student starts to move along the marked distance, and the timers start their stopwatches.

4 As the student passes each timer, the timer stops their stopwatch and notes the time. Collate all times for this first trial.

5 Repeat the trial for 2 or 3 other students. Repeat any trials where timing was incomplete or obviously wrong (for example, if the second student timer recorded a longer time than the third student).

DATA AND ANALYSIS

Record trials in the following table.

Distance travelled (m)	Trial 1 – Time taken (s)	Trial 2 – Time taken (s)	Trial 3 – Time taken (s)	Trial 4 – Time taken (s)
0.0	0.0	0.0	0.0	0.0
5.0				
10.0				
15.0				
20.0				
25.0				
30.0				

ISBN 978 1 4886 1935 9

PRACTICAL ACTIVITY 1.2

1 Graph position versus time for each trial, and compare the results.

2 Using the data table or position–time graph, construct a velocity–time graph for one of the trials.

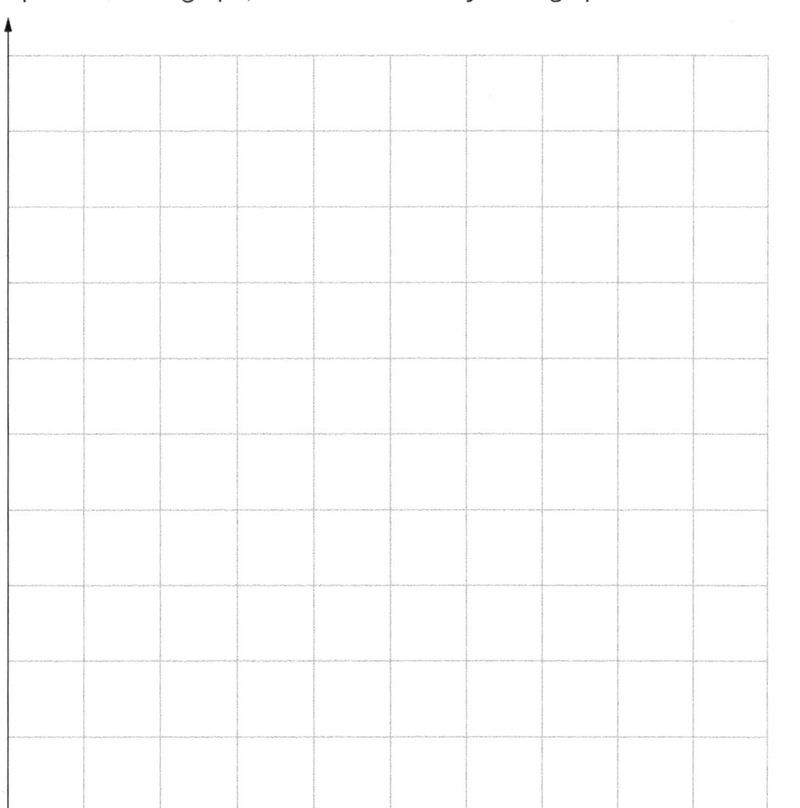

CONCLUSION

1 Describe how you can find velocity from the position–time graph.

2 Comment on the reliability of this means of measuring position, time and velocity.

3 What is the major source of error in this activity?

4 Suggest alternative methods that would improve the reliability of the results.

5 Consider a ticker timer that marked a constant distance interval rather than a constant time interval.

a Would this change the procedure by which you could calculate velocity?

b Would you expect this to give more accurate results?

RATING MY LEARNING	My understanding improved	Not confident ◄—► Very confident ○ ○ ○ ○ ○	I answered questions without help	Not confident ◄—► Very confident ○ ○ ○ ○ ○	I corrected my errors without help	Not confident ◄—► Very confident ○ ○ ○ ○ ○

 ISBN 978 1 4886 1935 9

PRACTICAL ACTIVITY 1.3

Analysing motion with a motion sensor

Suggested duration: 45 minutes

INTRODUCTION

When describing the motion of an object, it is essential to know where it is relative to a reference point, how fast and in what direction it is moving, and how it is accelerating (changing its rate of motion).

An ultrasonic motion sensor uses pulses of ultrasound that reflect from an object to determine the position of the object. As the object moves, the change in its position is measured many times each second. The change in position from moment to moment is expressed as a velocity ($m\,s^{-1}$). The change in velocity from moment to moment is expressed as an acceleration ($m\,s^{-2}$). The position of an object at a particular time can be plotted on a graph. You can also graph the velocity and acceleration of the object versus time.

A graph is a mathematical picture of the motion of an object. For this reason, it is important to understand how to interpret a graph of position, velocity or acceleration versus time. In this activity you will plot a graph of position in real time; that is, as the motion is happening.

MATERIALS

- motion sensor
- electronic measurement interface and software
- base and support rod

PURPOSE

To investigate the relationships between the motion of an object and a position–time graph for the object.

PROCEDURE

For this activity, you and a partner will be the objects in motion. The motion sensor will measure your position as you move in a straight line at different speeds. The software that supports your sensor will plot your motion on a graph of position and time.

Setting up the motion sensor

Set up the software for the following settings:

Graph: Position–time

Sampling rate: 10 samples/sec (10 Hz). Stop condition at 10 seconds if available.

Calibrate the sensor to allow for temperature if necessary. (You do not need to calibrate the sensor for temperatures around 20°C.)

> Consult the manuals for your electronic equipment or see your teacher if you are not sure how to set these values with your equipment and software.

1 Mount the sensor on a support rod so that it is aimed at your stomach when you are standing in front of it. Make sure that you can move at least 2 metres away from the sensor.

2 Position the device screen so you can see the screen while you move away from the sensor. Make sure the graph of position–time is visible and active.

3 When you are ready, stand in front of the sensor and click the Record button to begin recording data. Recording will begin almost immediately. The sensor will make a faint clicking noise. Your partner should not be able to see you while you are moving.

4 Watch the plot of your motion on the graph, and move backwards and forwards in a straight line so that the plot of your motion creates a relatively simple position versus time plot. Click the Stop button to finish measuring after about 10 seconds, if it does not stop automatically. Run #1 will appear in your Data list in the Experiment Setup window.

5 Now invite your partner to try and precisely match your graph. Your partner's trace can be set to appear directly over your own.

6 Try a more complex motion (but still in a straight line), or ask your partner to leave the room while you create the initial graph. Invite them back to try to match the graph of your movement without the opportunity of having seen your movement.

DATA AND ANALYSIS

1 Use the software, or the position–time data, to draw a velocity–time graph for your motion. Print a copy of your graph and insert it here.

2 Use your software, or the velocity–time data from step 1, to draw an acceleration–time graph. Insert a copy of your graph here.

 ISBN 978 1 4886 1935 9

CONCLUSION

1 **a** Comment on the ability to match another person's position–time graph.

b What were the main clues to establishing how they had originally moved?

2 Describe your motion using the information in one of your graphs.

RATING MY LEARNING	My understanding improved	Not confident ◀——▶ Very confident ○ ○ ○ ○ ○	I answered questions without help	Not confident ◀——▶ Very confident ○ ○ ○ ○ ○	I corrected my errors without help	Not confident ◀——▶ Very confident ○ ○ ○ ○ ○

PRACTICAL ACTIVITY 1.4

A reaction timer

Suggested duration: 45 minutes

INTRODUCTION

The acceleration due to gravity provides a simple, reliable and repeatable means of timing a person's reaction time when combined with an understanding of the equations of motion.

If t is the time taken by a person to react and catch a falling object that was initially at rest, then:

$$\vec{s} = \vec{u}t + \frac{1}{2}\vec{a}t^2$$

where:

$\vec{s}$ = distance fallen in metres

$\vec{u}$ = initial velocity in $m\,s^{-1}$

t = time of fall in seconds

$\vec{a}$ = acceleration in $m\,s^{-2}$.

MATERIALS

metre ruler

PURPOSE

To determine an individual's reaction time using a simple application of the equations of motion.

PROCEDURE

Repeated tests are required in this activity to ensure that the person is sufficiently unprepared for the time taken to be a true indication of reaction time.

1. Work in groups of at least two. The tester stands on a desk holding a metre ruler vertically a little out from their body so it can fall freely for the full length of the ruler. The person being tested holds thumb and forefinger around the end of the ruler closest to the ground without touching or impeding the initial fall of the ruler.
2. Once the person being tested is ready, let go of the ruler without alerting the person being tested. You may want to use some mild (but safe!) distraction to ensure the person being tested does not anticipate the fall. The person being tested grabs the ruler between thumb and forefinger, halting its fall as quickly as possible.
3. Record the distance that the ruler fell (in metres) and calculate the reaction time corresponding to this distance.
4. Repeat a number of times for both left and right hands, and average the results. Repeat for each person in the group.

DATA AND ANALYSIS

Rearrange the equation so that time t is on the left-hand side. Assume that gravitational acceleration is $9.8\,m\,s^{-2}$ downwards. Use this formula to calculate the reaction times, and enter the values into the following table.

Trial	Distance (m) right hand	Reaction time (s) right hand	Distance (m) left hand	Reaction time (s) left hand
1				
2				
3				
4				
Average				

ISBN 978 1 4886 1935 9

CONCLUSION

1 Collate results for everyone in the class. Comment on how your reaction time compares with the class average.

2 Based on the results for the whole class, is there any correlation between left- and right-handedness and the reaction time achieved with each hand?

3 Is there any correlation between reaction time and the types of sport played by students in the class? For example, fast reaction times are generally one key to success in sprints but not important for distance running.

4 A car is travelling at 60 km h^{-1}. In the event of a situation requiring emergency braking, how far will the car travel before the driver can actually start braking if their reaction time is 0.2 seconds?

RATING MY LEARNING	My understanding improved	Not confident ◄——► Very confident ○ ○ ○ ○ ○	I answered questions without help	Not confident ◄——► Very confident ○ ○ ○ ○ ○	I corrected my errors without help	Not confident ◄——► Very confident ○ ○ ○ ○ ○

DEPTH STUDY 1.1

Measuring height using kinematics

Suggested duration: 3.5–4 hours including data collection, analysis and report

INTRODUCTION

The study of kinematics includes the analysis of linear motion graphs, especially position–time, displacement–time, velocity–time and acceleration–time graphs. Because change in displacement is directly related to velocity and acceleration, measuring one or more of these quantities will enable you to calculate the height through which a lift has travelled.

This depth study requires you to conduct an investigation on a question you develop about how acceleration is related to displacement. You will research and analyse data and information, and communicate your findings in a written report including appropriate graphs and calculations to support your findings.

PURPOSE

To determine the height of a tower or multistorey building using kinematics data.

- Do not travel in service lifts. If you have to use a service lift for this activity, start recording and send your data recorder up and down by itself.
- Always seek permission before using a lift in a private building.
- Always let at least one other person know where you are going.

QUESTIONING AND PREDICTING

1 What measurements would best allow you to determine the change in height of a lift as you go up a building? Write a suitable inquiry question and hypothesis. Explain how these measurements could be used.

PLANNING AN INVESTIGATION

2 Consult with your teacher about what equipment is available to you for making and analysing your measurements. Research some alternatives that may be available to you. You should avoid using specialist equipment; look for ways to use equipment that is readily available to you. Write a list of your chosen materials.

ISBN 978 1 4886 1935 9

3 Describe your final method, using graphs to illustrate your intended calculations.

ANALYSING DATA AND INFORMATION

4 From your data, calculate the height of the building. Detail your analysis below. You may prefer to record some data in tables and print out any graphs you have used, with annotations to explain your analysis. Include any other relevant information, such as observations or measurements of the lift stops relative to the full height of the building. What other factors could influence your results?

Height of building:

Details of analysis:

COMMUNICATING

5 Prepare a written report, including appropriate graphs, tables and calculations.

Alternatively, present your investigation in a scientific poster format.

Whichever format you use, present the information in the following order:

- Summarise your initial predictions and measurements.
- List the equipment you used.
- Describe your method in a way that would let other students repeat your measurements and confirm your results.
- Present your data and result.

6 Discuss your result for the height of the building, considering the following:

- How does it compare with the stated height?
- What factors did you consider that gave you a good comparison?
- What factors that you did not consider may have affected your result?
- What error does your method introduce into your result? You may be able to calculate an estimate of the error.
- Does the device have a maximum acceleration that it is able to measure? Did this affect your result?

7 Could your method be applied to determining a distance in a horizontal direction? Write a suitable inquiry question and hypothesis. What limitations would need to be applied to measurement in a horizontal direction?

 ISBN 978 1 4886 1935 9

MODULE 1 • REVIEW QUESTIONS

Multiple choice

A new system of units has been proposed for the Republic of Utopia.

The fundamental units in this system are:

- the **hark**, the unit for distance (symbol h)
- the **furm**, the unit for acceleration (symbol f)
- the **worm**, the unit for time (symbol w)
- the **kard**, the unit for velocity (symbol k)

All other units are defined in terms of these fundamental units.

1 What is an alternative unit to the furm?

A hw

B hw^{-2}

C wk^{-1}

D kh

2 What unit is represented by the area under a velocity–time graph?

A hark

B furm

C worm

D kard

Consider the following quantities:

A 4 $m\,s^{-1}$

B 5 N downwards

C 45 kg

D 12 $m\,s^{-2}$ east

E 76 units to the right

F 64 km

G 45 $km\,h^{-1}$ at 337°T

3 Which of the quantities are correctly expressed as vectors?

4 Which of the quantities are correctly expressed as scalars?

5 In a 100 m race at the school swimming sports, Jeni completes four laps of the 25 metre pool in 72 s. Which of the following statements about the magnitudes of her speed and velocity is true?

A Her average speed and average velocity were both 0 $m\,s^{-1}$.

B Her average speed was 0 $m\,s^{-1}$, and her average velocity was about 1.4 $m\,s^{-1}$

C Her average speed was about 1.4 $m\,s^{-1}$, and her average velocity was 0 $m\,s^{-1}$.

D Her average speed and average velocity were both 1.4 $m\,s^{-1}$.

6 Starting from rest, a cyclist accelerates to 12 $m\,s^{-1}$ in a time of 4.0 s. What was her average acceleration during this time?

A 0 $m\,s^{-2}$

B 0.33 $m\,s^{-2}$

C 3.0 $m\,s^{-2}$

D 48 $m\,s^{-2}$

Short answer

7 A car is observed moving along a road at a steady speed of 60 $km\,h^{-1}$.

a How far, in kilometres, will it move in 1 minute?

b How long would it take to travel 20 km?

c How many metres does it move in 30 seconds?

8 A sports car takes 6 s to accelerate uniformly from rest to a speed of 30 m s^{-1}. It then maintains this velocity for 14 s before decelerating uniformly to a stop in a distance of 30 m.

a Draw a velocity–time graph illustrating this motion.

b How far does the car travel in its journey?

c How long does the journey take?

Extended response

9 A stationary car accelerates west for 100 m before reaching a constant velocity after 9.1 s.

a Calculate the constant velocity that the car reaches.

b Find the acceleration over this time.

The car then overtakes a bus travelling at 9 m s^{-1} in the same direction.

c What is the velocity of the car relative to the bus?

d Calculate the velocity of the bus relative to the car.

e In what direction does the bus appear to move in the car driver's frame of reference?

10 Describe, in terms of its velocity, the acceleration of a ball when bouncing off a wall. Using the wall as the frame of reference, what direction will the acceleration have?

 ISBN 978 1 4886 1935 9

MODULE

2 Dynamics

Outcomes

By the end of this module you will be able to:

- design and evaluate investigations in order to obtain primary and secondary data and information PH11-2
- select and process appropriate qualitative and quantitative data and information using a range of appropriate media PH11-4
- solve scientific problems using primary and secondary data, critical thinking skills and scientific processes PH11-6
- describe and explain events in terms of Newton's laws of motion, the law of conservation of momentum and the law of conservation of energy PH11-9

Content

FORCES

INQUIRY QUESTION **How are forces produced between objects and what effects do forces produce?**

By the end of this module you will be able to:

- using Newton's laws of motion, describe static and dynamic interactions between two or more objects and the changes that result from:
  - a contact force
  - a force mediated by fields
- explore the concept of net force and equilibrium in one-dimensional and simple two-dimensional contexts using: (ACSPH050) ICT N
  - algebraic addition
  - vector addition
  - vector addition by resolution into components
- solve problems or make quantitative predictions about resultant and component forces by applying the following relationships: ICT N
  - $\vec{F}_{AB} = -\vec{F}_{BA}$
  - $\vec{F}_x = \vec{F}\cos\theta$, $\vec{F}_y = \vec{F}\sin\theta$
- conduct a practical investigation to explain and predict the motion of objects on inclined planes (ACSPH098) CCT ICT

FORCES, ACCELERATION AND ENERGY

INQUIRY QUESTION **How can the motion of objects be explained and analysed?**

By the end of this module you will be able to:

- apply Newton's first two laws of motion to a variety of everyday situations, including both static and dynamic examples, and include the role played by friction (friction $= \mu\vec{F}_N$) (ACSPH063) CCT

Module 2 • Dynamics

- investigate, describe and analyse the acceleration of a single object subjected to a constant net force and relate the motion of the object to Newton's second law of motion through the use of: (ACSPH062, ACSPH063)
  - qualitative descriptions CCT
  - graphs and vectors ICT N
  - deriving relationships from graphical representations including $\vec{F} = m\vec{a}$ and relationships of uniformly accelerated motion ICT N
- apply the special case of conservation of mechanical energy to the quantitative analysis of motion involving: ICT N
  - work done and change in the kinetic energy of an object undergoing accelerated rectilinear motion in one dimension ($W = \vec{F}_{net}\vec{s}$)
  - changes in gravitational potential energy of an object in a uniform field ($\Delta U = m\vec{g}\Delta\vec{h}$)
- conduct investigations over a range of mechanical processes to analyse qualitatively and quantitatively the concept of average power ($P = \frac{\Delta E}{t}$, $P = \vec{F}\vec{v}$) including but not limited to: ICT N
  - uniformly accelerated rectilinear motion
  - objects raised against the force of gravity
  - work done against air resistance, rolling resistance and friction

MOMENTUM, ENERGY AND SIMPLE SYSTEMS

INQUIRY QUESTION **How is the motion of objects in a simple system dependent on the interaction between the objects?**

By the end of this module you will be able to:

- conduct an investigation to describe and analyse one-dimensional (collinear) and two-dimensional interactions of objects in simple closed systems (ACSPH064) CCT
- analyse qualitatively and predict, using the law of conservation of momentum ($\Sigma m\vec{v}_{before} = \Sigma m\vec{v}_{after}$) and, where appropriate, conservation of kinetic energy $\left(\Sigma\frac{1}{2}m\vec{v}^2_{before} = \Sigma\frac{1}{2}m\vec{v}^2_{after}\right)$, the results of interactions in elastic collisions (ACSPH066) ICT N
- investigate the relationship and analyse information obtained from graphical representations of force as a function of time
- evaluate the effects of forces involved in collisions and other interactions, and analyse quantitatively the interactions using the concept of impulse ($\Delta\vec{p} = \vec{F}\Delta t$) ICT N
- analyse and compare the momentum and kinetic energy of elastic and inelastic collisions (ACSPH066) ICT N

Key knowledge

Forces

Some forces act only on contact, while others can act at a distance. An example of a **contact force** is colliding billiard balls, where the contact force acts at the point when the balls collide. Examples of **non-contact forces**, or forces that act at a distance, include gravitation, magnetism and electrical force.

Force is a vector quantity, and its SI unit is the newton (N). The **net force** $\vec{F}_{net}$ acting on an object experiencing a number of forces acting simultaneously is given by the vector sum of all the individual forces.

$$\vec{F}_{net} = \vec{F}_1 + \vec{F}_2 + ... + \vec{F}_n$$

 A force can be thought of as a push or a pull.

Equilibrium exists when the vector sum of all forces acting on an object results in a zero net force. This is when the forces are balanced. An example of this is in a tug of war, when both teams apply a force in opposite directions but neither team moves. For a team to win it must pull with a greater force, which makes the forces unbalanced so that there is a net force on the rope in that team's direction. This concept is illustrated in Figure 2.1.

NEWTON'S FIRST LAW

Newton's first law can be written in many ways, but the two main ways are as follows:

1 An object will continue moving with a constant velocity unless an unbalanced force acts upon it.

2 A non-zero net force causes acceleration.

Newton's first law is also sometimes referred to as the Law of Inertia.

In other words, if the net force acting on an object is zero, then the object's motion will not change and it will maintain a constant velocity. This also applies if the object is at rest; its velocity is zero, and if no net force acts on it then some time later its velocity will still be zero.

This can be explained using the concept of terminal velocity, which is the speed a free-falling object reaches when its downward velocity is equal to the resistance of the medium (i.e. air resistance). Once an object reaches terminal velocity, the net force acting on it is equal to zero because the gravitational force equals the air resistance force, and it no longer accelerates.

Acceleration occurs when there is a change in velocity. If there is a net force acting on an object, its velocity will change and therefore it experiences acceleration.

This relates to Newton's first law, because the object will maintain its velocity (speed and direction) unless it is acted on by an unbalanced force.

 **Inertia** is the tendency of an object to resist a change in motion. An object with a large mass will have a large inertia; an object with a small mass will have a small inertia.

You can experience this phenomenon if you are standing in a train as it arrives at a station. If the train slows down quickly, you will stumble forward as though you have been pushed forwards. Because your body has inertia, your mass resists the change in motion and continues in the direction you were moving before the brakes were applied, so your body moves forward relative to the train.

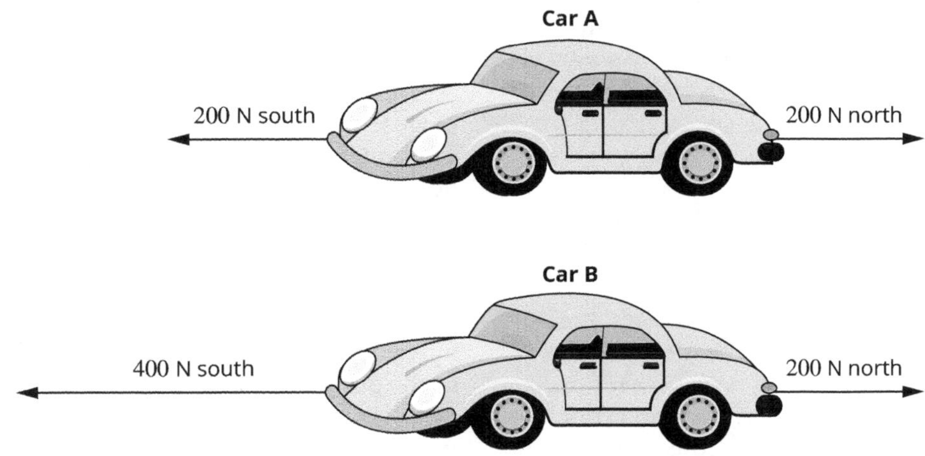

FIGURE 2.1 Balanced and unbalanced forces

NEWTON'S SECOND LAW

Newton's second law states that the acceleration of an object is directly proportional to the net force on the object and inversely proportional to the mass of the object.

Most people know this law unconsciously, because it requires more force to make a heavy object move than a light object. For example, it is easy to start an empty wheelbarrow moving, but it is much harder if the wheelbarrow is full.

This law can be expressed as a mathematical relationship: $\vec{F} = m\vec{a}$ where $\vec{F}$ = force (N), m = mass (kg) and $\vec{a}$ = acceleration ($m\,s^{-2}$).

For example, a model rocket with a mass of 0.050 kg is launched vertically upwards with an acceleration of $90\,m\,s^{-2}$. The net force acting on the rocket is then $\vec{F} = m\vec{a} = 0.050 \times 90 = 4.5\,N$ upwards.

However, this example considers only the net force acting on the object. Forces do not always act alone and this must be considered when doing calculations, because the overall effect of the forces depends on their direction. In the model rocket example, the net force is the result of the force produced upwards by the rocket's engine and the gravitational force acting downwards on the rocket. Because forces are vectors, they can be added or combined using the techniques discussed in Module 1.

It is a common misconception that gravity makes heavier objects fall faster than lighter objects. In fact, this is the result of gravity combined with air resistance. If air resistance is negligible, heavy objects fall at exactly the same rate as light objects.

This concept was discussed in relation to terminal velocity in Newton's first law. If a flat piece of A4 paper and a scrunched up piece of A4 paper are dropped together, the scrunched-up paper will hit the ground first because there is less air resistance acting on it.

When an object is moving through air, air resistance is a force that acts in the opposite direction to the direction of motion. It therefore reduces the acceleration of falling objects.

Another misconception is that the gravitational force is the same for all objects; in fact it is the gravitational acceleration that is the same. Different gravitational forces act on different masses to produce the same acceleration. A large mass has more inertia than a small mass, so a greater gravitational force (**weight**) is needed to accelerate it at the same rate.

NEWTON'S THIRD LAW

For every action (force) there is an equal and opposite reaction (force). This is known as **Newton's third law**.

This law is the most widely known of Newton's three laws, but it is often misunderstood. It basically means that in every interaction (including non-contact forces), there is a pair of forces acting on the two interacting objects. The action force and reaction force are acting on different objects, so it is important to recognise that they should not be added together.

If the action force is labelled systematically, the reaction force can be described by reversing the label of the action force. An example of this is shown in Figure 2.2, where the action force is the force on the bat by the ball, and the reaction force is the force on the ball by the bat.

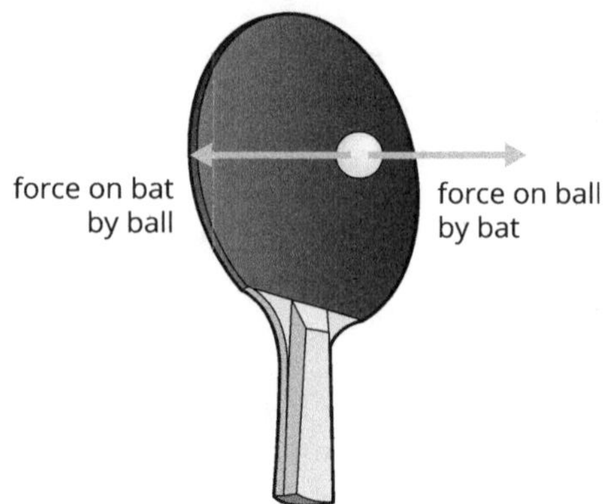

FIGURE 2.2 The action–reaction pair of a ball hitting a table tennis bat is an example of Newton's third law.

The action and reaction forces are equal and opposite even when the masses of the colliding objects are very different. This is the main reason why this law is often misunderstood. In the example in Figure 2.2, the force applied to the bat by the ball is the same as the force applied to the ball by the bat, but the ball goes flying and the bat moves only slightly. The individual forces making up Newton's third law pair act on different masses to cause different accelerations, as shown by Newton's second law. The mass of the ball in Figure 2.2 is a lot smaller than the mass of the bat, and so it accelerates faster.

TABLE 2.1 Action and reaction force pairs.

Action force	Reaction force
A cannonball is shot out of the cannon.	The cannon recoils with the same force. Because it is heavier, the cannon is affected less.
Hot gases are forced out of the engine of a rocket.	The gases exert an equal and opposite force on the rocket. This causes the rocket to accelerate.

A force that is often confused with an action–reaction pair is the **normal force** represented by $\vec{F}_N$. The normal force is always equal and opposite to the weight force of an object on a surface. An example of this is a coffee cup sitting on a bench. Even though there is a gravitational force pulling it towards the Earth, there is an equal and opposite force holding it in position (Figure 2.3a). Even though this seems like an example of Newton's third Law because the forces are both acting on the coffee cup, it cannot be a pair of forces. Instead, the normal force (i.e. the upwards force on the cup by the table) is the Newton's third law pair with the downwards force on the table by the cup.

If an object is on an inclined plane, the normal force applied to the object by the plane decreases as the angle of the plane increases. Gravitational force always acts vertically downwards. So if the plane is inclined, the normal force must balance the vertical component of the gravitational force vector. This situation is represented in Figure 2.3b.

 ISBN 978 1 4886 1935 9

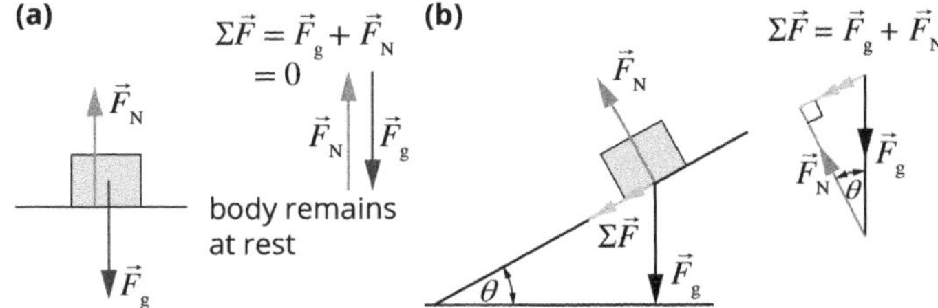

FIGURE 2.3 Normal force on (a) a horizontal plane, and (b) on an inclined plane

Forces, acceleration and energy

FORCES AND FRICTION

Friction is a force that opposes movement. It has very important uses on Earth yet it is often discussed in negative terms. Since friction is a resistive force that produces heat energy, this consequently causes energy inefficiency. However, this resistive quality of friction is also necessary in many applications such as holding items or preventing slipping and sliding. When considering surfaces in contact with one another, there are two different types of friction to consider; static and kinetic friction.

The frictional force between stationary surfaces is called static friction, and can be represented by $\vec{F}_s$.

Static friction is due to the interlocking of the irregularities between two surfaces and the forces of attraction between the atoms and molecules. Consider a force being applied to an object on a surface. The object remains stationary until enough force is applied to overcome friction. This maximum frictional force is the static friction. Once the object begins to slide, a much lower force than the static friction force is needed to keep the object moving at a constant velocity.

The frictional force between sliding surfaces is called kinetic friction, and can be represented by $\vec{F}_k$.

Static friction is greater than kinetic friction, because before one surface can move across the other, the interlocking bonds must be broken. This requires a much greater force because once there is motion between the surfaces, the bonds cannot re-form.

The amount of friction between two surfaces can be calculated using the following equation:

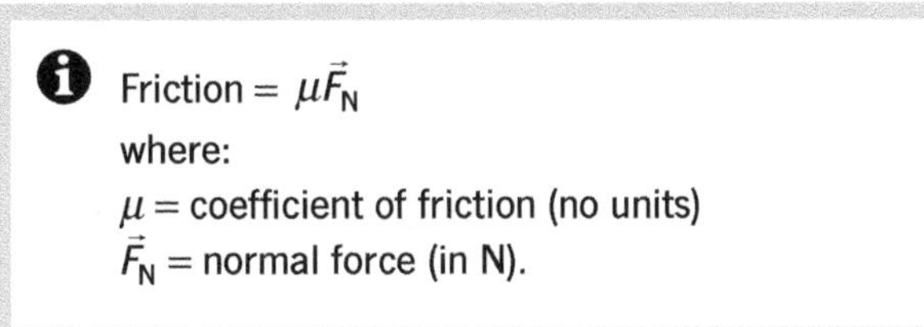

Friction = $\mu \vec{F}_N$

where:

μ = coefficient of friction (no units)

$\vec{F}_N$ = normal force (in N).

The coefficient of friction, μ, is a dimensionless constant which depends on the nature of the surfaces and whether the situation involves static friction (μ_s) or kinetic friction (μ_k).

Consider a person pushing a steel box that weighs 200 N across a steel platform. If the coefficient of kinetic friction between the box and platform is 0.6, the friction formula can be used to determine the force of friction acting on the box:

$$\begin{aligned} \text{friction} &= \mu \vec{F}_N \\ &= 0.6 \times 200 \\ &= 120\,\text{N in the opposite direction to the motion of the box.} \end{aligned}$$

Friction often causes moving objects to decelerate. The amount of deceleration can be calculated using Newton's second law. In the example above, the deceleration of the steel box would be:

$$\begin{aligned} \vec{a} &= \frac{\vec{F}_{net}}{m} \\ &= \frac{120}{20} \\ &= 6\,\text{m}\,\text{s}^{-2} \end{aligned}$$

WORK

Energy and work have many meanings in everyday life, but in physics they have very precise meanings. It is important to understand what these terms mean and how they are related.

Energy is the capacity to cause a change.

A simple example of energy causing a change is baking a fruit cake. The heat energy supplied by the oven causes a physical change. The mixture of flour, water, sugar, eggs, fruit and spices is transformed from a soggy mass into a firm cake.

Energy is a scalar quantity; it has magnitude but not direction. The SI unit for energy is joules (J).

Energy is conserved in a reaction. It can be transferred or transformed, but it cannot be created or destroyed. An example of this is the process of photosynthesis. Light energy from the Sun is transformed by plants into chemical energy, which the plants use for cellular processes.

There are many different forms of energy. These can be broadly classified as either **kinetic** (associated with motion) or **potential** (associated with position). Some examples are shown in Table 2.2.

TABLE 2.2 Examples of kinetic and potential energy.

Kinetic energy	Potential energy
electrical energy: movement of electrons	chemical energy: relative positions of atoms
sound energy: vibrations of particles	gravitational energy: position in gravitational field

In physics, work cannot be done without energy, and vice versa. Work is equal to the change in energy, and therefore its unit is also the joule. Work is done when:

- energy is transferred or transformed
- a force causes an object to be displaced.

When you pick up a box, a force is applied to the box and the box is lifted, which means work has been done on the box. **Work** is the product of net force and displacement: $W = \vec{F}_{net}\vec{s}$. The unit for work is joules and the units for force and displacement are newtons and metres, so from this formula you can see that 1 J is equivalent to 1 N m.

When a force produces no displacement, no work is done. In Figure 2.4, no work is being done on the box even though there are energy transformations going on inside the person's body to hold the box in position. It is important to remember that the energy of the box is not changing, so no work is being done on the box to keep it at a constant height. Table 2.3 gives some examples of situations where work is done, and work is not done.

FIGURE 2.4 No work is done on a box when you are holding it at a constant height.

TABLE 2.3 Examples of situations where work is, and is not, done.

Situation	Is work done?
A car is pushed along a road.	Work is being done, because a force acts over a distance.
You are sitting on a chair.	No work is being done, because there is no displacement.
A tap is dripping.	Work is being done, because a gravitational force is acting on the drips over a distance.

ENERGY CHANGES

All moving objects have kinetic energy. The kinetic energy of an object is equal to the work required to accelerate the object from rest to its final velocity.

The kinetic energy of an object is given by the equation:

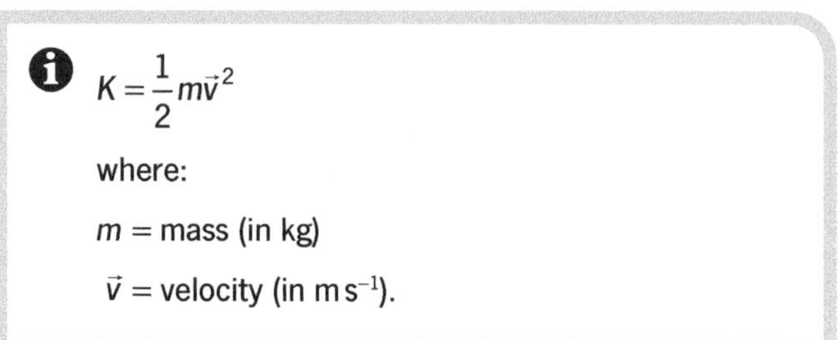

$$K = \frac{1}{2}m\vec{v}^2$$

where:

m = mass (in kg)

$\vec{v}$ = velocity (in m s^{-1}).

The work–energy theorem defines work as a *change* in kinetic energy:

$$W = \frac{1}{2}m\vec{v}^2 - \frac{1}{2}m\vec{u}^2 = \Delta K$$

where:

W is work (in J)

m is mass (in kg)

$\vec{u}$ is initial velocity (in m s^{-1})

$\vec{v}$ is final velocity (in m s^{-1}).

In theory all work is converted into kinetic energy, but in reality energy is lost from the system because of friction.

It is also important to note that the definitions for kinetic energy and change in kinetic energy can be derived entirely from known concepts such as work, Newton's second law and the equations of motion.

When considering potential energy, only gravitational potential energy will be discussed because it is one of the most important forms of potential energy. Gravitational potential energy is the energy an object has because of its position in a gravitational field. The higher the object is displaced, the larger its gravitational potential energy.

The change in gravitational potential energy of an object, ΔU, is given by the equation:

$$\Delta U = m\vec{g}\Delta\vec{h}$$

where:

m = mass (in kg)

$\vec{g}$ = gravitational field strength (9.8 m s^{-2}, or 9.8 N kg^{-1}, downwards on Earth)

$\Delta\vec{h}$ = change in height (in m).

Gravitational potential energy is calculated relative to a zero potential energy reference level. On Earth this is usually the ground or sea level.

MECHANICAL ENERGY AND POWER

Mechanical energy is the sum of the potential and kinetic energies of an object. In other words, it is the energy associated with the object's motion or position, or both.

Considering kinetic energy and gravitational potential energy, mechanical energy can be calculated as follows:

$$E_m = K + U = \frac{1}{2}m\vec{v}^2 + m\vec{g}\Delta\vec{h}$$

This formula is very useful in energy transformations. The principle of conservation of energy states that the total mechanical energy in a system remains constant if the only forces acting are conservative forces.

As discussed above in relation to work, in all real systems non-conservative forces like friction are present, but they are mostly considered to have negligible values or they can be used to help identify other important energy transformations. For example, when a ball bounces, some mechanical energy is transformed into heat and sound, which is why the ball will eventually stop bouncing. This can be compared to an object being dropped through air; in that situation only a small

ISBN 978 1 4886 1935 9

amount of its energy will be converted to heat and sound, so this small effect can be considered negligible for falling objects. Therefore, mechanical energy is conserved in falling objects.

Conservation of mechanical energy can be used to predict outcomes in a range of situations involving gravity and motion.

For example, Figure 2.5 shows a 70 kg skateboarder at the top of a 2 m tall half-pipe. Initially, when the skateboarder is at the top of the pipe, the total mechanical energy is equal to the gravitational potential energy stored at that height.

Therefore, at the top:

$$E_m = K + U = 0 + m\vec{g}\Delta\vec{h} = 70 \times 9.8 \times 2 = 1372\,\text{J}$$

This value is constant because of the conservation of energy in falling objects, so the total mechanical energy at the bottom of the slope is the same. When the skateboarder reaches the bottom his gravitational potential energy is 0 J, which means his kinetic energy is 1372 J.

To extend this scenario, the final speed of the skateboarder at the bottom of the half-pipe could be calculated. This occurs when all of the gravitational potential energy is converted into kinetic energy, so:

$$\begin{aligned} K &= U \\ \frac{1}{2}m\vec{v}^2 &= m\vec{g}\Delta\vec{h} \\ \vec{v}^2 &= 2\vec{g}\Delta\vec{h} \\ \vec{v} &= \sqrt{2\vec{g}\Delta\vec{h}} \\ &= \sqrt{2\times9.8\times2} \\ &= 6.3\,\text{m s}^{-1} \end{aligned}$$

FIGURE 2.5 A skateboarder at the top of a 2 m tall half-pipe

Power is a measure of the rate at which work is done:

$$P = \frac{\Delta E}{t}$$

where:

P = power (in W)

ΔE = energy transferred or transformed (in J)

t = time taken (in s).

The unit of power is the watt (W). Much like energy, the word power means different things in everyday language. However, these meanings all relate to the idea that the more powerful something is, the more work it can do in a certain amount of time.

Often it is convenient to calculate the average power. For example, when pushing a box across the floor where there is friction involved, a force is needed to keep the box sliding at a constant speed. The power required to keep an object moving at a constant speed can be calculated from the product of the force applied and its average speed: $P = \vec{F}\vec{v}$.

The two types of friction discussed earlier were static and sliding friction, but these frictions only occur when surfaces are sliding across each other or about to slide across each other. There are other types of friction that are experienced every day, such as rolling resistance and air resistance.

Momentum, energy and simple systems

CONSERVATION OF MOMENTUM

Momentum is the product of an object's mass and velocity. It is a vector quantity and is calculated using the equation:

$$\vec{p} = m\vec{v}$$

where:

m = mass (in kg)

$\vec{v}$ = velocity (in m s^{-1}).

An object in motion has momentum, and if this momentum changes the object must experience a force. Force is equal to the rate of change of momentum. The momentum of a system will be conserved so that $\sum \vec{p}_{\text{before}} = \sum \vec{p}_{\text{after}}$.

TABLE 2.4 Situations where the conservation of momentum can be applied

Situation	Change in momentum
Two objects collide and remain separate.	$\sum \vec{p}_{\text{before}} = \sum \vec{p}_{\text{after}}$ $m_1\vec{u}_1 + m_2\vec{u}_2 = m_1\vec{v}_1 + m_2\vec{v}_2$
Two objects collide and combine together.	$\sum \vec{p}_{\text{before}} = \sum \vec{p}_{\text{after}}$ $m_1\vec{u}_1 + m_2\vec{u}_2 = m_3\vec{v}_3$
One object breaks apart into two objects in an explosive collision.	$\sum \vec{p}_{\text{before}} = \sum \vec{p}_{\text{after}}$ $m_1\vec{u}_1 = m_2\vec{v}_2 + m_3\vec{v}_3$

TABLE 2.5 Different types of collisions

Situation	Example
Two objects collide and remain separate.	The white ball hits a red ball in a game of snooker.
Two objects collide and combine together.	A footballer running and tackling another player.
One object breaks apart into separate objects in an explosive collision.	A clay target is hit during a skeet-shooting event.

The law of conservation of kinetic energy is written as:

$$\sum \frac{1}{2} m\vec{v}^2_{\text{before}} = \sum \frac{1}{2} m\vec{v}^2_{\text{after}}$$

In **elastic collisions** the kinetic energy of the system is conserved. In an **inelastic collision** kinetic energy is converted into other types of energy like thermal and sound.

MOMENTUM TRANSFER

Change or transfer in momentum, $\Delta\vec{p}$, is also known as **impulse**. It is a vector quantity. A change or transfer in momentum occurs when an object changes its velocity.

The equation for impulse is:

$$\Delta\vec{p} = m\vec{v} - m\vec{u} = m(\vec{v} - \vec{u})$$

Change in velocity is determined by:

$$\Delta\vec{v} = \text{final velocity} - \text{initial velocity}$$

This is shown in Figure 2.6a. Using the head-to-tail method and Pythagoras' theorem, the change in velocity can then be calculated (Figure 2.6b). Using the vector addition skills in Module 1, the impulse can then be found by multiplying the change in velocity by the mass of the object.

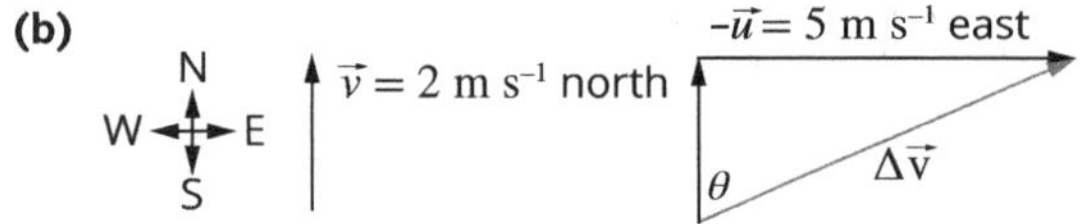

FIGURE 2.6 Calculating impulse with vector addition

MOMENTUM AND NET FORCE

Newton's second law describes the relationship between impulse, force and the period of time during which the force acts:

$$\Delta\vec{p} = m(\vec{v} - \vec{u}) = m\vec{a}\Delta t$$

$$\therefore \Delta\vec{p} = \vec{F}\Delta t$$

There are a few important things about this relationship:

- The same mass changing its velocity by the same amount will have a constant change in momentum or impulse.
- The faster a mass changes its velocity, the greater the force required to change the velocity in that period of time.
- An object colliding with a hard surface stops in a short time, but an object colliding with a soft surface takes a longer time to stop.
- The period of time taken to stop is proportional to the force involved. A shorter time means a greater force, while a longer time means a smaller force.

This concept can be used to explain the different effects of impacts. For example, it explains why an egg dropped on a hard floor breaks, but an egg dropped on a pillow does not. In both cases m and $\Delta\vec{v}$ are the same. However, the pillow greatly increases the time the egg takes to stop, so the force acting on the egg is much less.

In most problems it is assumed that the force remains constant throughout the collision, but in reality this is not always the case. Forces can change during a collision. The impulse over a period of time can be found by calculating the area under the line on a force versus time graph (Figure 2.7).

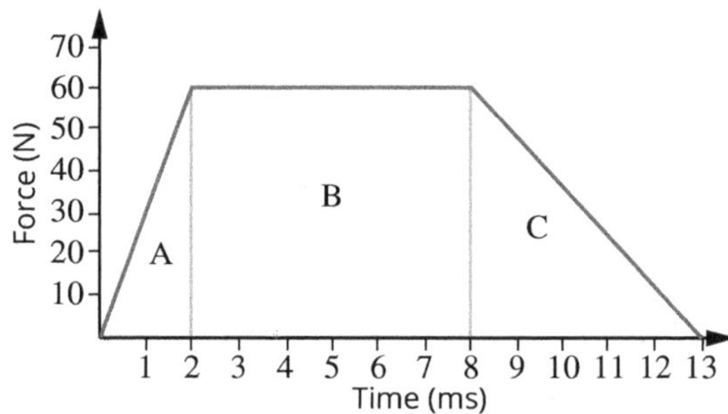

FIGURE 2.7 The area under a force–time graph gives the total impulse of the event.

$$\begin{aligned}
\text{area} &= A + B + C \\
&= \frac{1}{2}(b_A \times h_A) + (b_B \times h_B) + \frac{1}{2}(b_C \times h_C) \\
&= \frac{1}{2} \times (2.0 \times 10^{-3} \times 60) + (6.0 \times 10^{-3} \times 60) + \\
&\quad \frac{1}{2} \times (5.0 \times 10^{-3} \times 60) \\
&= 0.060 + 0.36 + 0.15 \\
&= 0.57 \\
\text{impulse} &= 0.57\,\text{kg m s}^{-1}
\end{aligned}$$

 ISBN 978 1 4886 1935 9

WORKSHEET 2.1

Knowledge review—dynamics

1 When you are standing up on a bus, it is a good idea to make sure you are hanging onto something. Using Newton's first law, explain why this is the case.

2 Starting from rest, a 1200 kg car accelerates at 2.0 m s^{-1} for 4.0 s.

a What is its speed after 4.0 s?

b What force is required to achieve this acceleration?

c How much kinetic energy does it have after 4.0 s?

d How far does it travel in that time?

3 A 5.0 kg model sailboat is moving with a constant acceleration, as shown below.

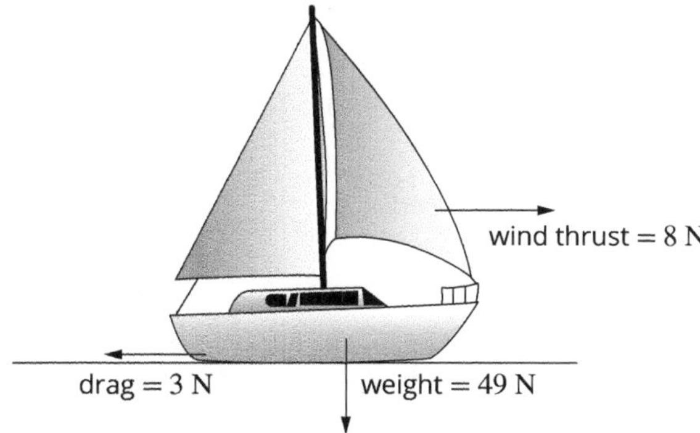

There is a force missing from the diagram. What is it, and what is its magnitude?

4 When a ball bounces on a hard floor, the 'action' is the force of the ball downwards on the floor. What is the 'reaction'?

WORKSHEET 2.2

Which Newton?

In each of the following questions, identify the relevant law of motion that applies, and state the appropriate answer.

1 What happens to the velocity of an object if it never experiences an unbalanced force?

Newton's law: ☐ 1st ☐ 2nd ☐ 3rd

2 A horse that is hitched to a stationary trolley begins to pull on the trolley. If the force exerted by the trolley on the horse is always equal in size and opposite in direction to the force exerted by the horse on the trolley, how can the horse move the trolley?

Newton's law: ☐ 1st ☐ 2nd ☐ 3rd

3 We know from experience that the harder we throw a ball (apply more force), the faster it will move (greater initial velocity resulting from acceleration). If you throw a 1 kg ball as hard as you can, and it leaves your hand at $20\,\mathrm{m\,s^{-1}}$, how fast do you think a 5 kg ball would leave your hand when you throw it as hard as you can?

Newton's law: ☐ 1st ☐ 2nd ☐ 3rd

4 An object's resistance to change in motion is called inertia. What property of matter gives an object inertia? Give an example of something with a relatively large amount of inertia, and something else with a relatively small amount of inertia.

Newton's law: ☐ 1st ☐ 2nd ☐ 3rd

5 If a rocket's engine produces a constant force, will the rocket's acceleration when it is launched be the same as the acceleration just before the fuel is completely used up? Explain your answer.

Newton's law: ☐ 1st ☐ 2nd ☐ 3rd

6 An astronaut is about 20 m from her space station when the small rocket thrusters she uses to move around run out of fuel. If she is carrying a few tools and can also remove the rocket thruster, what should she do to get back to the space station? Explain your reasoning.

Newton's law: ☐ 1st ☐ 2nd ☐ 3rd

7 A force sensor is used to measure the force applied to drag a 1.5 kg brick across a table while a motion sensor is used to measure the brick's acceleration. Several trials are conducted, and the slope of the force versus acceleration graph is about 2.0 in each trial, which is lower than might be expected. What might explain the difference?

Newton's law: ☐ 1st ☐ 2nd ☐ 3rd

8 What would happen to a ball thrown into deep space where there were no forces acting on it? Describe its motion during the time you are in contact with it and then after release.

Newton's law: ☐ 1st ☐ 2nd ☐ 3rd

RATING MY LEARNING	My understanding improved	Not confident ◄——► Very confident ○ ○ ○ ○ ○	I answered questions without help	Not confident ◄——► Very confident ○ ○ ○ ○ ○	I corrected my errors without help	Not confident ◄——► Very confident ○ ○ ○ ○ ○

 ISBN 978 1 4886 1935 9

WORKSHEET 2.3

Understanding Newton's third law

The following is a snippet from an article covering some of the misconceptions of Newton's third law. Imagine someone is holding a ball, as shown below.

'What is the force that is equal and opposite the gravitational field force exerted by the Earth on the ball?' said the physics teacher.

'The force exerted by the hand on the ball.' said the class.

Physics teacher replied, 'Let's say this ball is in free fall, what forces are on it?'

Now only the gravitational field force exerted by the Earth on the ball is left, so what is the force that is equal and opposite of this? Turns out that the force that is equal and opposite the gravitational field force exerted by the Earth on the ball is the gravitational field force exerted by the ball on the Earth. The ball accelerates toward the Earth, and the Earth accelerates toward the ball. However, because of the Earth's mass compared to the ball's mass, the ball accelerates at a much larger rate. The forces are always directly toward each other, and are always equal and opposite in direction. The equal and opposite forces are acting on different objects altogether.

(Source: M. J. Hughes, 2002. How I misunderstood Newton's third law. *The Physics Teacher* volume 40, number 6, pages 381–382.)

1 Explain one common misconception associated with Newton's third law.

2 Compare Newton's third law to Newton's first and second laws and explain how they are connected.

3 Describe a situation that involves all three laws.

4 For the situation you described in Question 3, use reasonable estimates of the mass, starting velocities and acceleration to estimate the forces involved in the interaction between the objects.

RATING MY LEARNING	My understanding improved	Not confident ◄——► Very confident ○ ○ ○ ○ ○	I answered questions without help	Not confident ◄——► Very confident ○ ○ ○ ○ ○	I corrected my errors without help	Not confident ◄——► Very confident ○ ○ ○ ○ ○

WORKSHEET 2.4

Friction is the rub

1 Describe, using diagrams and worded explanations, an experience involving friction. What factors determine the frictional force between objects?

2 If you push on a heavy box that is at rest on a surface, you must exert some force to start its motion. However, once the box is sliding, you can apply a smaller force to maintain its motion. Why?

3 Suppose you are driving a car at high speed. A driving instructor would tell you that you should avoid applying the brakes rapidly when you want to stop in the shortest possible distance. In fact, you should apply the brakes gradually and perhaps even briefly release the brakes at times. (Newer cars have antilock brakes that avoid this problem.) Using what you have learnt about static and kinetic friction, explain why this is good advice.

4 A farmer places a large crate in the metal tray of a ute, but does not tie it down. As the ute accelerates forward, the crate remains at rest relative to the tray of the ute. What force allows the crate to accelerate forward with the ute?

5 If the farmer in Question 4 applied the brakes rapidly, what would happen to the crate? Why?

6 A trolley with wooden wheels is launched up an inclined plane. After going up the plane, it stops and then slides back down to its starting position. The coefficient of kinetic friction between the trolley's wheels and the plane is 0.25. Will the time for the trip up the plane be less than or greater than the time for the trolley to go down the inclined plane? Explain your answer.

RATING MY LEARNING	My understanding improved	Not confident ◄—► Very confident ○ ○ ○ ○ ○	I answered questions without help	Not confident ◄—► Very confident ○ ○ ○ ○ ○	I corrected my errors without help	Not confident ◄—► Very confident ○ ○ ○ ○ ○

 ISBN 978 1 4886 1935 9

WORKSHEET 2.5

Working against the force (of gravity)

Circle the best option in each of the following multiple-choice questions:

1 Mechanical work is related to which pair of variables?

A Distance and time
B Distance and force
C Distance and velocity
D Distance and acceleration

2 A raven carries a stick from the ground to its nest in a tree. Complete this statement: 'The work done on the stick is directly related to the stick's ...?'

A gravitational potential energy
B elastic potential energy
C kinetic energy
D none of the above

3 A ball is thrown straight up into the air and returns to be caught by the thrower. We know that gravity has been acting on the ball the entire time, but the ball has the same kinetic energy when it is caught as when it was released. Can you explain this apparent discrepancy?

4 A 1000 kg rocket is launched vertically upwards at a constant acceleration. When it reaches an altitude of 15 km it is travelling at 900 m s^{-1}.

a How much work was done by the rocket?

b Using the work done by the rocket, what is the magnitude of its acceleration?

c How long did the rocket take to reach 15 km altitude?

5 A seal with a mass of 120 kg slides straight down a 35° slope of ice for 70 metres. At the end of the slope it is travelling at 30 m s^{-1}. Assuming the slope is frictionless, what was the seal's initial velocity?

RATING MY LEARNING	My understanding improved	Not confident ◄——► Very confident ○ ○ ○ ○ ○	I answered questions without help	Not confident ◄——► Very confident ○ ○ ○ ○ ○	I corrected my errors without help	Not confident ◄——► Very confident ○ ○ ○ ○ ○

WORKSHEET 2.6

Real-time graphs—estimating area

Finding the area of a field

A farmer must know the area that has been prepared for a crop in order to know exactly how much seed and fertiliser are required. If this is not known exactly then either the crop will not get planted correctly or there will be wastage. Both are expensive mistakes.

For a rectangular field, the calculation is relatively simple but not every paddock is rectangular. How would a farmer measure the area of a paddock with an irregular border such as the river shown in the following diagram? There is a way without resorting to integration.

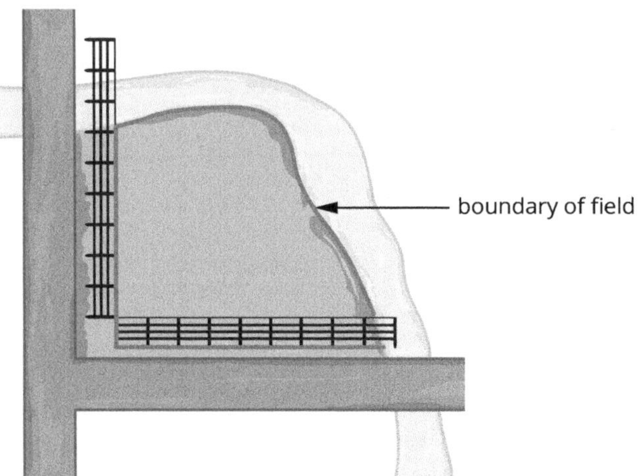

1 To obtain a first-order approximation of the area, multiply the widest distance by the longest distance. If the fence posts in the diagram are 2 metres apart, what is a first-order approximation of the area of the paddock between the fences and the river?

2 A second-order approximation uses known points to make smaller rectangles. The fence posts along the sides of the paddock provide useful known points. Using each fence post at the bottom of the paddock in the diagram as a known point, rule lines straight up to the edge of the river, parallel to the fence on the left.

3 From the point where the vertical lines intersect the lower bank of the river, complete a rectangle by drawing a horizontal line from that point to the left until it meets the fence line or previous parallel line, whichever is closer. It is valid if your horizontal lines extend into the river.

4 Calculate the area of each individual rectangle you have drawn and add the areas together. (Round each rectangle height to the nearest fence post.) How does this result compare to your first-order approximation?

5 What could you do to make the approximation better? Describe how a third-order approximation could be made of the paddock area.

 ISBN 978 1 4886 1935 9

Impulse from a force–time graph

Impulse is an important concept when investigating change in momentum and the forces generated in collisions. This might involve a ball hitting the ground, a rocket engine firing, or something more commonly associated with the word 'collision', such as a car crash.

In the real world, force–time graphs are rarely straight lines. The diagram below shows a typical graph drawn using data collected when a dynamics trolley hit a force sensor.

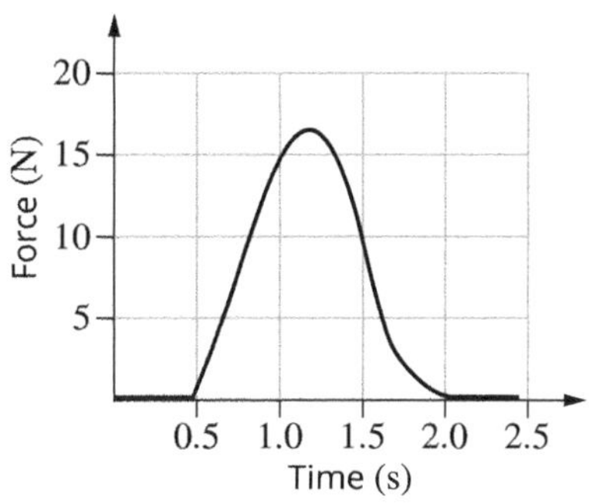

6 Using the units shown on the force–time graph, estimate the total impulse during the collision using both a first-order and a second-order approximation.

7 Given the definition of a first-order approximation, is it possible for a first-order approximation to be smaller than the actual area under the curve? Explain.

8 Is it possible for a second-order approximation to be the same as the actual area under a curve? Justify your answer.

9 What quantity would the area under the following curve represent?

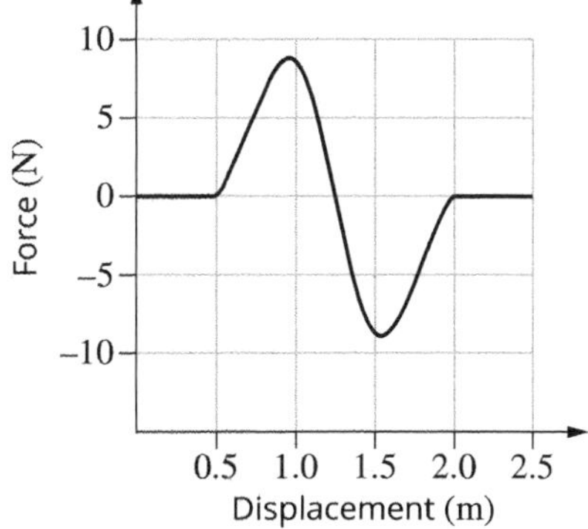

10 Estimate the area under this curve.

RATING MY LEARNING	My understanding improved	Not confident ◄——► Very confident ○ ○ ○ ○ ○	I answered questions without help	Not confident ◄——► Very confident ○ ○ ○ ○ ○	I corrected my errors without help	Not confident 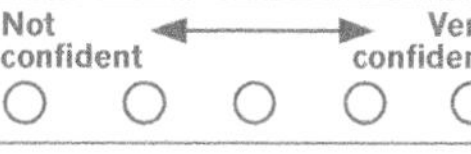 Very confident ○ ○ ○ ○ ○

WORKSHEET 2.7

Engine power

Most internal combustion engines have a well-defined 'power band'. The power band is defined as the range of engine speeds (in terms of engine rpm, or revolutions per minute) over which the engine delivers its greatest power.

Typically, a power band for a four-stroke engine would range from 3500 rpm to 4500 rpm. It is possible for an engine to operate at speeds below or above the power band, but such operation will seriously affect fuel economy and may even damage the engine. Other types of internal combustion engines such as two-stroke, diesel or rotary may exhibit quite different characteristics.

The graph below depicts the relationship between the traction force a car can apply to the speed at which it can travel. Notice that there are 5 curves, each for a different gear in a 5-speed car. Having a range of different gears allows the engine to operate optimally at many different speeds.

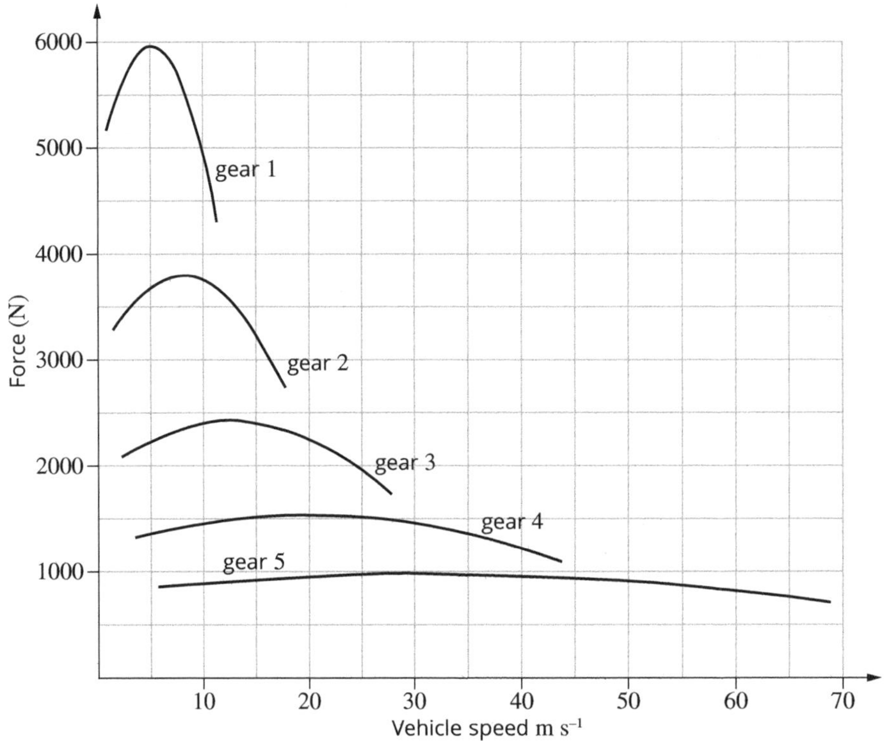

1 Using the data in the graph, fill in the table below for each gear for peak force (F_{peak}), peak velocity (v_{peak}), minimum velocity (v_{min}) and maximum velocity (v_{max}).

Gear	$\vec{F}_{peak}$ (N)	$\vec{v}_{peak}$ (m s^{-1})	$\vec{v}_{min}$ (m s^{-1})	$\vec{v}_{max}$ (m s^{-1})	P (W)
1st					
2nd					
3rd					
4th					
5th					

 ISBN 978 1 4886 1935 9

2 Plot a graph of $\vec{F}_{peak}$ versus $\vec{v}_{peak}$. What relationship is suggested by the graph?

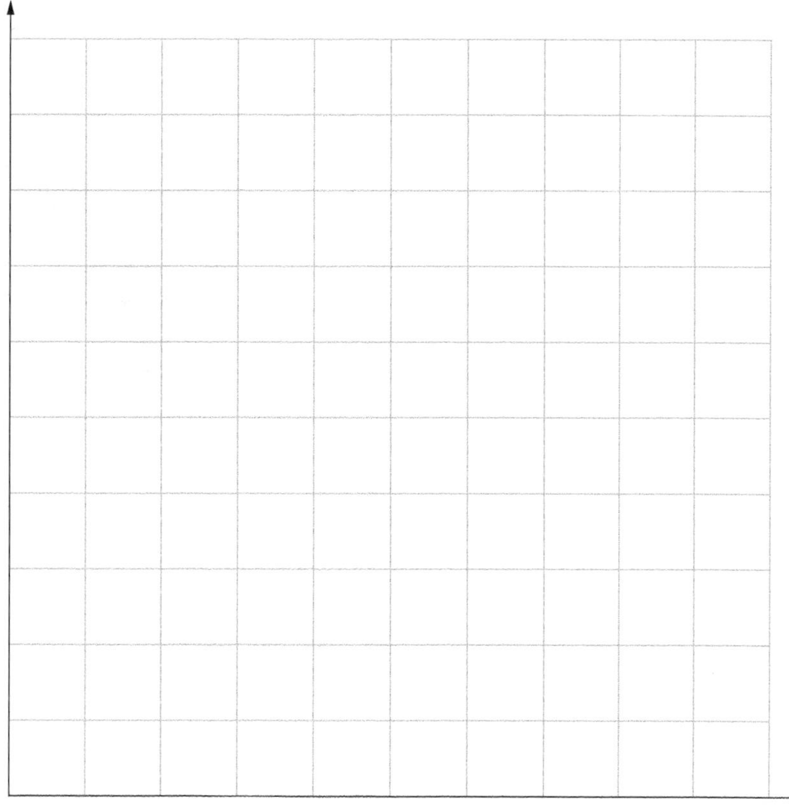

3 Recall that power is the rate of doing work or $P = \frac{\Delta E}{\Delta t} = \vec{F}\vec{v}$. Calculate the optimum power for each of the five gears assuming that the optimum power is given by $\vec{F}_{peak}$. Add this data to the table above. What do you notice?

4 Assuming that the car in this example has a mass of 1200 kg, at 5 m s^{-1} in 1st gear how much traction can the car apply?

5 What is the car's maximum acceleration at that point?

6 When this car tows an 800 kg caravan, its performance changes markedly. The additional mass will reduce the possible acceleration. Ignoring air resistance, what acceleration is now possible at 5 m s^{-1} in first gear?

7 How will the caravan affect the maximum acceleration at 30 m s^{-1}? What will it do to the maximum speed attainable?

RATING MY LEARNING	My understanding improved	Not confident ◀——▶ Very confident ○ ○ ○ ○ ○	I answered questions without help	Not confident ◀——▶ Very confident ○ ○ ○ ○ ○	I corrected my errors without help	Not confident ◀——▶ Very confident ○ ○ ○ ○ ○

ISBN 978 1 4886 1935 9

WORKSHEET 2.8

Human earthquake

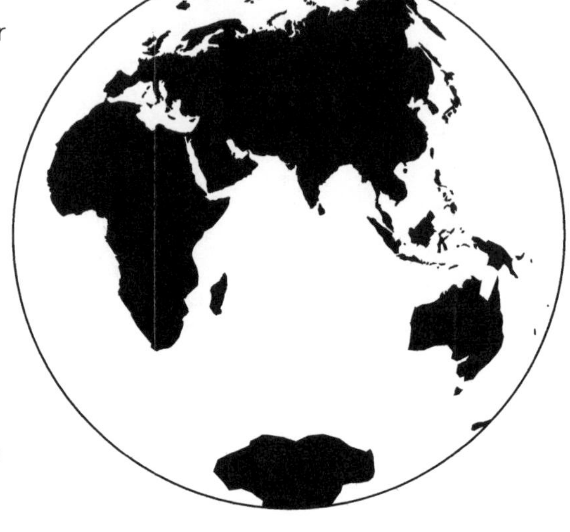

From time to time it has been suggested that, if millions of people in the same country (e.g. China) all jumped down from a height at exactly the same time, an earthquake could occur. The idea is that their combined gravitational potential energy would be converted to kinetic energy, which would then be transferred to the Earth.

Your task is to determine whether this idea has any merit. Use reasonable estimates of the quantities involved, including the impact time for people landing on the ground, to determine:

1 The initial gravitational potential if the whole population of China were standing 2.0 metres above the ground.

2 The total kinetic energy at the time of impact if they all landed at the same time.

3 The impulse generated on impact, and hence the force, based on a reasonable estimate of stopping time.

4 The average mechanical power during this impact.

5 The velocity at which the Earth would move after the impact, assuming all of the energy from the jump is transferred to the Earth.

The following figures will assist in determining your estimate.

Mass of the Earth: 5.97×10^{24} kg

Population of China: 1400 million

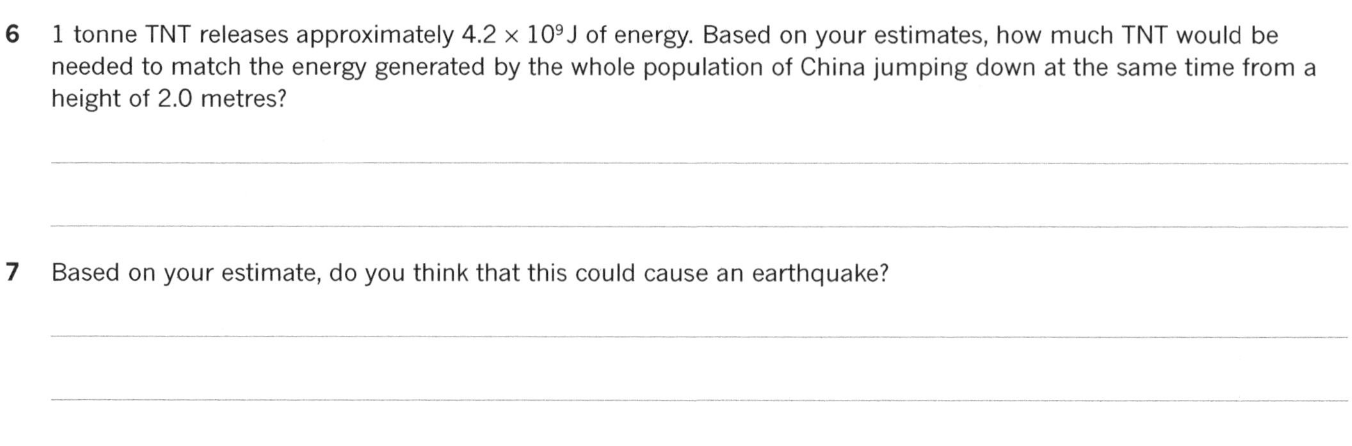

6 1 tonne TNT releases approximately 4.2×10^{9} J of energy. Based on your estimates, how much TNT would be needed to match the energy generated by the whole population of China jumping down at the same time from a height of 2.0 metres?

7 Based on your estimate, do you think that this could cause an earthquake?

RATING MY LEARNING	My understanding improved	Not confident ◄——► Very confident ○ ○ ○ ○ ○	I answered questions without help	Not confident ◄——► Very confident ○ ○ ○ ○ ○	I corrected my errors without help	Not confident ◄——► Very confident ○ ○ ○ ○ ○

ISBN 978 1 4886 1935 9

WORKSHEET 2.9

Literacy review—the language of forces

In the following statements, fill in the blanks from the list of words below. You can use a word more than once.

first	second	weight
acceleration	displacement	energy
force	kinetic energy	mass
mechanical work	potential energy	work–energy theorem
conserved	conservation of momentum	conservation of energy
before	after	conservation of mass
during	momentum	vectors
collisions	explosion	internal forces
external forces	elastic	inelastic
greater	less	the same
velocity	half	double
proportional	friction	gravitational potential
partial	total	work

1 Newton's ________________ law predicts the following relationship between ________________, force, and mass: The acceleration of an object is directly ________________ to the net force and will always be in the same direction as the net ________________. Acceleration is inversely proportional to the ________________ of the object, meaning that more massive objects have less acceleration if subjected to the same net force.

2 An object subject to an unbalanced force will experience a ________________. The work done by that unbalanced force is equal to the component of the ________________ in the direction of the object's displacement multiplied by the magnitude of the displacement. If the ________________ imparted to the object by the force is converted entirely into kinetic energy, the work done by the unbalanced force is equal to the change in the object's ________________. This concept is known as the ____________________.

3 The momentum of an object equals its ________________ multiplied by its ________________. When two objects collide, the total momentum of the system is ________________ as long as there are no outside forces acting on the system. This concept is known as the law of ________________, which states that the total momentum of the system ________________ an event (such as a collision) is the same as the momentum ________________ the event.

4 There are several types of ________________, which are events where ________________ is often conserved. ________________ collisions occur when both ________________ and ________________ are conserved, while ________________ collisions occur when ________________ is not conserved. Another type of momentum-related event is the ________________, where one object may split into two or more separate objects.

5 In an ideal frictionless environment, a trolley poised at the top of a hill has ________________ ________________ energy. As the trolley begins to roll down the hill, this ________________ is transformed into ________________ energy. When the trolley reaches the bottom of the hill the ________________ energy of the system consists entirely of ________________. In the real world, some of this energy would be lost to ________________ in the form of heat.

RATING MY LEARNING	My understanding improved	Not confident ◄──► Very confident ○ ○ ○ ○ ○	I answered questions without help	Not confident ◄──► Very confident ○ ○ ○ ○ ○	I corrected my errors without help	Not confident ◄──► Very confident ○ ○ ○ ○ ○

WORKSHEET 2.10

Thinking about my learning

After completing Module 2: Dynamics, you should be able to describe, explain and apply the relevant scientific ideas. You should be able to work with data to interpret, analyse and evaluate it.

1 The table below lists the key knowledge covered in this module. Read each and reflect on how well you understand each concept. Rate your learning by shading the circle that corresponds to your level of understanding for each concept. It may be helpful to use colour as a visual representation. For example:

- green—very confident
- orange—in the middle
- red—starting to develop.

Concept focus	**Rate my learning**				
	Starting to develop ◄——► Very confident				
Newton's first law of motion	○	○	○	○	○
Newton's second law of motion	○	○	○	○	○
Newton's third law of motion	○	○	○	○	○
Motion of objects on inclined planes	○	○	○	○	○
Friction	○	○	○	○	○
Conservation of mechanical energy	○	○	○	○	○
Work	○	○	○	○	○
Average power	○	○	○	○	○
Conservation of momentum	○	○	○	○	○
Elastic and inelastic collisions	○	○	○	○	○
Impulse	○	○	○	○	○

2 Consider where you have shaded in one of the first three circles. List any specific concepts that you found challenging in this module.

3 Write down two different strategies that you will apply to help improve your understanding of these concepts.

ISBN 978 1 4886 1935 9

PRACTICAL ACTIVITY 2.1

Acceleration down an incline

Suggested duration: 45 minutes

INTRODUCTION

A trolley on an incline will roll down the incline because it is pulled by gravity, as shown in the diagram below.

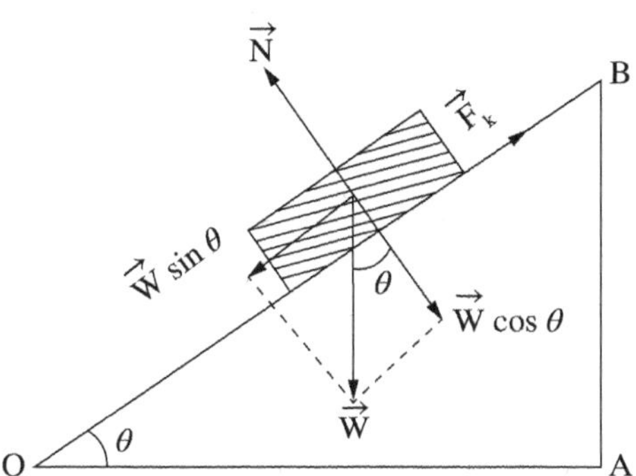

Gravitational acceleration is directed straight down. The component of this acceleration that is parallel to the inclined surface is $m\vec{g}\sin\theta$. So this is the net acceleration of the trolley down the incline, ignoring friction.

To measure the acceleration, the trolley will start from rest, and you will measure the time (t) it takes for it to travel a certain displacement ($\vec{s}$).

Since $\vec{s} = \vec{u}t + \frac{1}{2}\vec{a}t^2$ and $\vec{u} = 0$, the acceleration provided by gravity down the plane can be found using:

$$\vec{a} = \frac{2\vec{s}}{t^2}$$

A plot of acceleration versus $\sin\theta$ will be a straight line with a gradient equal to the acceleration due to gravity ($\vec{g}$).

MATERIALS

- dynamics trolley
- masses
- stopwatch
- dynamics track or board at least 1 metre long
- end stop, book or block (to stop the trolley)
- protractor
- metre ruler
- graph paper
- cardboard
- a ticker timer can be used for this activity

PURPOSE

To study the effect of the angle of an incline on the acceleration of an object.

You could increase the accuracy of the activity by measuring the acceleration of the trolley with a motion sensor or photogates, and graph the motion directly to a computer for analysis.

Fix a piece of cardboard (about the size of a playing card) across the front an airtrack glider to act as a reflector for the motion sensor signal. The dynamics trolley has enough frontal area to reflect the signal without modification. If the sensitivity of the sensor can be adjusted, set it to 'near' to reduce the effect of stray echoes.

If your dynamics trolley has an integrated motion encoder, use the following settings:

Sample rate: 10–20 Hz

Graph: velocity versus time

Calculate: acceleration for each trial

Manual entry: graph of acceleration versus $\sin\theta$

PRACTICAL ACTIVITY 2.1

PROCEDURE

1 Set up the track as shown in the diagram, raising the end of the track where the trolley is to start by about 20 cm. Record the angle the ramp makes with the horizontal.

2 Pull the trolley up to the top of the track and record the displacement of this initial position from the end stop where the trolley will complete its run. Make sure you record the distance from the front of the trolley.

3 Release the trolley from rest and use the stopwatch to time how long it takes the trolley to hit the end stop. Repeat this measurement 5 times, with a different person doing the timing each time. Record all the values in the table below.

4 Lower the end of the track by 2 cm, measure the new angle to the horizontal, and then measure the time for another 5 runs.

5 Repeat the experiment for a total of 7 angles, lowering the track by 2 cm for each new angle.

DATA AND ANALYSIS

1 Calculate the average time the trolley took for each angle.

2 Calculate the acceleration using the displacement and times recorded with your data.

3 Calculate $\sin\theta$ from the angle for each of the heights.

Height (m)	$\vec{s}$ (m)	θ (°)	Time 1 (s)	Time 2 (s)	Time 3 (s)	Time 4 (s)	Time 5 (s)	Average time (s)	$\vec{a}$ (m s^{-2})	$\sin\theta$

4 Plot a graph of acceleration versus $\sin\theta$ in the space provided. Draw a straight line of best fit to your data and calculate its slope in order to find $\vec{g}$.

ISBN 978 1 4886 1935 9

CONCLUSION

1 Comment on the relationship you found between acceleration and the angle of incline.

2 What value did you obtain for $\vec{g}$? How does this compare with the accepted value for your location? Can you explain any discrepancy?

3 Does your reaction time cause a greater percentage error for higher or lower angles?

4 If you doubled the mass of the trolley, how would the results be affected? Try it to test your hypothesis.

5 Which method provides the most accurate measure of the angle of the track: (a) calculation from length and height, or (b) measurement with the protractor? Justify your answer.

RATING MY LEARNING	My understanding improved	Not confident ◄——► Very confident ○ ○ ○ ○ ○	I answered questions without help	Not confident ◄——► Very confident ○ ○ ○ ○ ○	I corrected my errors without help	Not confident ◄——► Very confident ○ ○ ○ ○ ○

PRACTICAL ACTIVITY 2.2

Newton's second law, part 1

Suggested duration: 20+ minutes

INTRODUCTION

Newton's second law states that the acceleration $\vec{a}$ of an object is directly proportional to the net force $\Sigma\vec{F}$ acting on it, and indirectly proportional to its mass m.

Written as an equation, Newton's second law becomes $\vec{F} = m\vec{a}$.

This is a vector equation where the direction of the acceleration is in the same direction as the net force.

MATERIALS

- dynamics trolley with plunger
- masses
- dynamics track (or simple board that can be levelled)
- end stop (or set up the track against a solid wall)

PURPOSE

To investigate and confirm the dependence of acceleration on force and mass.

PROCEDURE

1 Level the track by adjusting the levelling feet until a trolley placed on the track does not move. (If there are no levelling feet, use cardboard or books for levelling.)

To perform each of the following trials, attach the spring plunger to the trolley and place the trolley at rest at the end of the track with the plunger against the end stop (or convenient classroom wall). Then release the plunger by pressing the button on the trolley with a ruler. Observe the resulting acceleration. This will be a qualitative measurement; that is, you will be recording observations rather than making direct measurements.

Vary the force

3 Perform the first trial with the spring plunger set to the first possible position (the least compression), and then do two more trials increasing the force applied to the trolley by increasing the compression of the spring plunger.

4 Repeat each test at least twice to confirm your observations.

Vary the mass

5 For these trials, always set the spring plunger to the maximum. Observe the relative accelerations of the trolley alone and the trolley with one mass bar in it. If additional masses are available, use them to increase the mass for additional trials.

6 Repeat the test for each mass (and the trolley with no additional mass) at least twice to confirm your observations.

DATA AND ANALYSIS

1 Based on your observations, and with reference to them, does the acceleration increase or decrease as the force is increased? Explain with reference to your tests.

ISBN 978 1 4886 1935 9

2 Based on your observations, does the acceleration increase or decrease as the mass is increased? Explain with reference to your tests.

CONCLUSION

1 Do your results confirm Newton's second law? Comment on any apparent discrepancies.

2 What errors will affect the results of this experiment? How could the effect of these errors be reduced?

RATING MY LEARNING	My understanding improved	Not confident ◄—► Very confident ○ ○ ○ ○ ○	I answered questions without help	Not confident ◄—► Very confident ○ ○ ○ ○ ○	I corrected my errors without help	Not confident ◄—► Very confident ○ ○ ○ ○ ○

PRACTICAL ACTIVITY 2.3

Newton's second law, part 2

Suggested duration: 40+ minutes

INTRODUCTION

According to Newton's second law, $\vec{F} = m\vec{a}$ or $\vec{F} \propto \vec{a}$. $\vec{F}$ is the net force acting on an object of mass m and $\vec{a}$ is the resulting acceleration of the object.

For a trolley of mass m_1 on a horizontal track, attached to a mass m_2 by a string over a pulley, the net force $\vec{F}$ on the entire system (trolley plus hanging mass) is the weight of hanging mass, $\vec{F} = m_2\vec{g}$, assuming that friction is negligible.

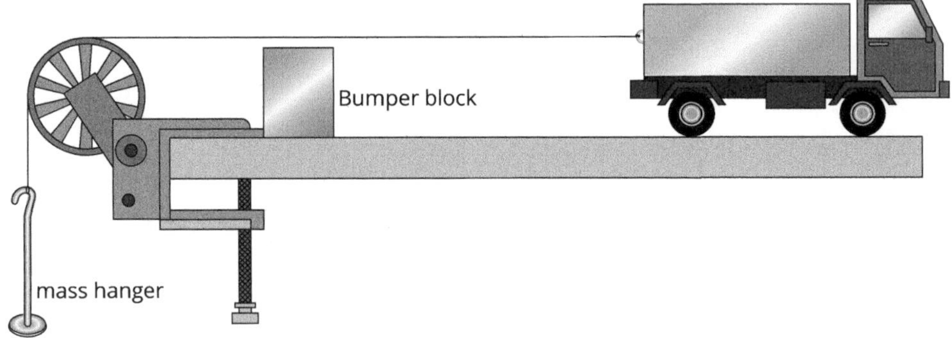

According to Newton's second law, this net force should be equal to $m\vec{a}$, where m is the total mass that is being accelerated, which in this case is $m_1 + m_2$. This experiment will check to see if $m_2\vec{g}$ is equal to $(m_1 + m_2)\vec{a}$ when friction is ignored.

To find the acceleration, the trolley will start from rest and the time (t) it takes for it to travel a certain distance ($\vec{s}$) will be measured. Then since $\vec{s} = \vec{u}t + \frac{1}{2}\vec{a}t^2$ and $\vec{u} = 0$, the acceleration can be found by rearranging in terms of $\vec{a}$ so that $\vec{a} = \frac{2\vec{s}}{t^2}$

MATERIALS

- dynamics trolley
- pulley with clamp
- string
- stopwatch
- mass balance
- chalk
- dynamics track (or simple board that can be levelled)
- mass hanger and mass set
- wooden or metal stopping block
- metre ruler

PURPOSE

To verify Newton's second law, $\vec{F} = m\vec{a}$.

PROCEDURE

1 Level the track by adjusting the levelling feet until a trolley placed on the track does not move. (If there are no levelling feet, use cardboard or books for levelling.)

2 Use the balance to find the mass of the trolley and record the value in the space provided under 'Data and analysis'.

3 Attach the pulley to the end of the track as shown in the diagram. Attach a string to the end of the trolley and place the trolley on the track. Tie the mass hanger onto the other end of the string and pass it over the pulley, so that it hangs over the end of the bench. The string must be just long enough so that the trolley hits the stopping block before the mass hanger reaches the floor.

4 Pull the trolley back until the mass hanger reaches the pulley. Record the position of the trolley and mark it on the track with the chalk. This will be the release position for all the trials. Make some test runs to determine how much mass is required on the mass hanger so that the trolley takes about 2 seconds to complete the run. Because of your reaction time, a short run time will result in a large percentage error. On the other hand, if the trolley moves too slowly then friction may result in a significant percentage error. Record the mass of the total hanging mass (hanger plus masses).

 ISBN 978 1 4886 1935 9

PRACTICAL ACTIVITY 2.3

If available, consider substituting electronic timing methods such as motion sensors, photogates or motion encoders to reduce measurement errors.

5 Place the trolley against the end stop on the pulley end of the track, and record the final position of the trolley.

6 Measure the time it takes for the trolley to travel from start to the pulley end position, and repeat at least 3 times Record these values in the table provided.

7 Increase the mass of the trolley and repeat the procedure.

DATA AND ANALYSIS

1 Calculate the total distance travelled by taking the difference between the initial and final positions of the trolley:

$\vec{s}$ = ____________________ m.

Total hanging mass m_2 = ____________________ kg.

Initial mass of the trolley m_1 = ____________________ kg.

2 Complete the data table for the remaining values calculated from your initial measurements. To determine a measure of the reliability of the comparison, calculate the percentage difference for each trial mass.

Percentage difference = $\%\Delta = \frac{\vec{F}_{net} - (m_1 + m_2)\vec{a}}{\vec{F}_{net}} \times \frac{100}{1}$

Trial	mass m_1 (kg)	$m_1 + m_2$ (kg)	t_1 (s)	t_2 (s)	t_3 (s)	t_4 (s)	t_{ave} (s)	$\vec{a} = \frac{2\vec{s}}{t^2}$ ($m\,s^{-2}$)	$(m_1 + m_2)\vec{a}$ (N)	$\vec{F}_{net}$ ($m_2\vec{g}$) (N)	% change
1											
2											
3											
4											
5											

CONCLUSION

1 To what extent do your observations, measurements and calculations confirm, or otherwise, Newton's second law? Comment on the reliability of your results with reference to the percentage difference.

2 Why is the mass in $\vec{F} = m\vec{a}$ not just equal to the mass of the trolley?

3 When calculating the force on the trolley using mass times gravity, why isn't the mass of the trolley included?

RATING MY LEARNING	My understanding improved	Not confident ◄——► Very confident ○ ○ ○ ○ ○	I answered questions without help	Not confident ◄——► Very confident ○ ○ ○ ○ ○	I corrected my errors without help	Not confident ◄——► Very confident ○ ○ ○ ○ ○

PRACTICAL ACTIVITY 2.4

Work, energy and power

Suggested duration: 45 minutes

INTRODUCTION

Mechanical work is a common aspect of everyday life. It is defined as a net force applied to an object over a distance that causes the object's kinetic energy to change. For example, when an apple falls from a branch, gravity does work, causing a change in the apple's kinetic energy. That is:

$$W = \vec{F}_{\text{net}}\vec{s} = \frac{1}{2}m\vec{v}^2_{\text{final}} - \frac{1}{2}m\vec{v}^2_{\text{initial}}$$

In this practical activity the motion of a simple trolley and track system is used to help understand this relationship.

MATERIALS

- data collection system
- motion sensor
- force sensor
- dynamics trolley and track (or ramp)
- end stop or heavy mass
- pulley and clamp
- mass and hangers
- balance (only one required for the class)
- 1.5 metre length of non-stretch string or braided physics string

PURPOSE

To investigate the relationship between the net force applied to an object over a distance and the resulting change in the object's kinetic energy.

PROCEDURE

1 Place the dynamics track on a lab table with one end hanging slightly over the edge of the table.

2 Place the dynamics trolley in the middle of the track, and then adjust the track level until the trolley does not roll.

3 Mount the pulley and clamp to the end of the track hanging over the edge of the table. Be sure the area below the pulley is clear to allow the mass and hanger to fall cleanly.

4 Mount the end stop to the track (or sit the heavy mass on the track) to prevent the trolley from colliding with the pulley.

5 Mount the motion sensor to the other end of the track with the sensing element pointing toward the pulley. Be sure the motion sensor switch is set on the trolley's position (if available) to avoid reflections from objects outside the trolley/track system.

6 Mount the force sensor to the top of the trolley. You will need to adjust the height of the pulley so that the top of the pulley is even with the force sensor hook when the trolley is on the track.

7 Measure the mass of the trolley and sensor together and record the value in the data table.

8 Connect the force sensor and motion sensor to the data collection system.

9 Place the dynamics trolley and force sensor on the track, and press the zero button on the force sensor to remove the force.

10 In the data collection software, display a force versus position graph (refer to your system's manual for instructions).

11 Set up the trolley and mass hanger:
- Hold the trolley on the track at the position you will be releasing it (10 cm away from the motion sensor).
- Attach a length of string, slightly longer than the distance from the force sensor to just below the pulley, to the force sensor hook. Attach the mass hanger to the opposite end of the string, hanging over the pulley.
- Place 50 grams on the hanger. Remember to add the mass of the hanger to the 50 grams when doing calculations.

ISBN 978 1 4886 1935 9

12 Start data recording in your data collection system.

13 Release the trolley, and let it run the length of the track.

14 Stop data recording just before the trolley collides with the end stop or the mass hits the floor, whichever comes first.

15 If time permits, repeat the experiment using different masses.

DATA AND ANALYSIS

Print out a copy of the force–position graph from the data analysis system and paste it in the space provided or draw your graph over the grid.

Using the graph, find the work done by calculating the area under the force–position graph. Check your answer against the value calculated by the software. Record the value of work done in the table below.

1 Why is it important that the force sensor hook is at the same height as the top of the pulley?

2 Why is a force–position graph used for this investigation?

ISBN 978 1 4886 1935 9

3 Explain the method you used to find the area under the force–position graph.

4 From your data collection system, find the initial and final velocities of the data run.

Initial velocity:

Final velocity:

5 What is the significance of the initial and final velocities to work done?

6 Draw the graph of velocity–time in the space provided below. You can also print and paste in the graph from your data analysis system if the feature is available.

7 Using the definition of power, how could you find the average power of this system from the collected data?

 ISBN 978 1 4886 1935 9

8 Using the results calculated above, complete the table below for each mass tested.

	Mass 1	Mass 2	Mass 3
Mass of trolley + force sensor + mass (kg)			
Work done (N m)			
$\vec{v}_{initial}$ (m s^{-1})			
$\vec{v}_{final}$ (m s^{-1})			
$K_{initial}$ (J) $= \frac{1}{2}m\vec{v}^2_{initial}$			
K_{final} (J) $= \frac{1}{2}m\vec{v}^2_{final}$			
$\Delta K = K_{(final)} - K_{(initial)}$ (J)			
Average power (W)			

CONCLUSION

1 Calculate the percentage difference between the change in work done and the change in kinetic energy. Using this value as a measure of the comparison between these two values, comment on how well, or otherwise, your experiment confirmed that $\Delta W = \Delta K$.

You may need to use the equation $\frac{\Delta W}{\Delta K} \times 100\%$.

2 What are some possible reasons for any difference?

3 What are some other forms of mechanical energy?

4 Why do you think these other forms of mechanical energy can be ignored in this investigation?

5 A light has been designed which uses a falling mass to power it. As the potential energy is transformed into kinetic energy, the light turns on. The light itself needs 0.1 W to turn on. How does this compare to the average power of the system for each of the masses?

RATING MY LEARNING

My understanding improved — Not confident  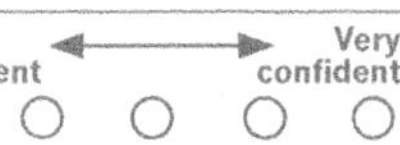 Very confident

I answered questions without help — Not confident 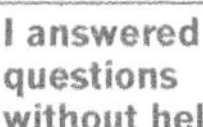 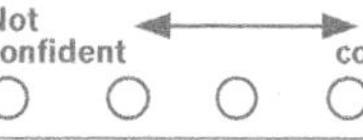 Very confident

I corrected my errors without help — Not confident ○ ○ ○ ○ ○ Very confident

ISBN 978 1 4886 1935 9

PRACTICAL ACTIVITY 2.5

Conservation of momentum

Suggested duration: 30+ minutes

INTRODUCTION

When two trolleys push away from each other and no net force exists, the total momentum of both trolleys is conserved:

$$\vec{p} = m_1\vec{v}_1 - m_2\vec{v}_2 = 0$$

Because the system is initially at rest, the final momentum of the two trolleys must be equal in magnitude and opposite in direction, so the resulting total momentum of the system is zero.

The ratio of the final speeds of the trolleys is therefore equal to the ratio of the masses of the trolleys:

$$m_1\vec{v}_1 - m_2\vec{v}_2 = 0$$
$$m_1\vec{v}_1 = m_2\vec{v}_2$$
$$\frac{\vec{v}_1}{\vec{v}_2} = \frac{m_1}{m_2}$$

In this investigation, the starting point for both trolleys is their position at rest, so that the two trolleys reach the end of the track simultaneously. To determine the speed, the distance travelled can be measured, since the time travelled by each trolley is the same.

This eliminates one potential source of error from the experiment, and the ratio of the distances is equal to the ratio of the masses:

$$\frac{\Delta\vec{s}_1}{\Delta\vec{s}_2} = \frac{m_1}{m_2}$$

MATERIALS

- 2 dynamics trolleys
- masses
- dynamics track, or board at least 1.5 metres long
- metre ruler
- mass balance
- chalk

PURPOSE

To demonstrate conservation of momentum for two trolleys pushing away from each other.

 This activity can also use an airtrack and gliders.

PROCEDURE

1 Level the track by adjusting the levelling feet until a trolley placed on the track does not move. (If there are no levelling feet, use cardboard or books for levelling.)

For each of the following cases, place the two trolleys against each other with the plunger of one trolley pushed completely in and latched in its maximum position.

2 Use a pencil to push the plunger release button of one trolley, and watch the two trolleys move in opposite directions to the ends of the track. Experiment with different starting positions until the two trolleys reach their end of the track at the same time. Mark this starting position with chalk to ensure it is exactly the same for each trial.

3 Weigh the two trolleys and record the masses and the starting position.

 ISBN 978 1 4886 1935 9

4 Carry out the following trials. Each trial should be run at least three times, and the distance travelled by each trolley recorded for each run.

Trial 1: Trolleys of equal mass — use two trolleys without mass bars.

Trial 2: Trolleys of unequal mass — put one mass bar in one trolley, and none in the other.

Trial 3: Trolleys of unequal mass — put two mass bars in one trolley, and none in the other.

Trial 4: Trolleys of unequal mass — put two mass bars in one trolley, and one mass bar in the other.

DATA AND ANALYSIS

1 For each of the trials, measure the distances travelled from the starting position to the respective ends of the track for each trolley and record them in the data table below.

2 Calculate the ratio of the distances travelled and record your results in the data table.

3 Calculate the ratio of the masses and record this in the data table.

Trial	$\vec{s}_{\text{trolley 1}}$ (m)	$\vec{s}_{\text{trolley 2}}$ (m)	$\frac{\vec{s}_1}{\vec{s}_2}$	$\frac{m_1}{m_2}$
Trial 1, Run 1				
Trial 1, Run 2				
Trial 1, Run 3				
Trial 2, Run 1				
Trial 2, Run 2				
Trial 2, Run 3				
Trial 3, Run 1				
Trial 3, Run 2				
Trial 3, Run 3				
Trial 4, Run 1				
Trial 4, Run 2				
Trial 4, Run 3				

CONCLUSION

1 Does the ratio of the displacements equal the ratio of the masses in each of the cases? In other words, is momentum conserved? Comment on your results.

2 When trolleys with unequal masses push away from each other, which trolley has more momentum?

3 When the trolleys with unequal masses push away from each other, which trolley has more kinetic energy?

4 Does the starting position depend on which trolley has its plunger set? Why?

5 No units are shown in the data table for the ratios of distance and of mass. What should the units be for each ratio?

RATING MY LEARNING	My understanding improved	Not confident ◄——► Very confident ○ ○ ○ ○ ○	I answered questions without help	Not confident ◄——► Very confident ○ ○ ○ ○ ○	I corrected my errors without help	Not confident ◄——► Very confident ○ ○ ○ ○ ○

 ISBN 978 1 4886 1935 9

PRACTICAL ACTIVITY 2.6

Conservation of energy

Suggested duration: 40+ minutes

INTRODUCTION

The gravitational potential energy gained by a trolley as it climbs an incline is given by:

$$\Delta U_g = m\vec{g}\Delta\vec{h}$$

where:

m is the mass of the trolley (in kg)

$\vec{g}$ is the gravitational acceleration ($9.8\,m\,s^{-2}$ downwards on Earth)

$\vec{h}$ is the vertical height the trolley is raised (in m).

The height is given by:

$$\vec{h} = \vec{s}\sin\theta$$

where:

$\vec{s}$ is the displacement along the incline (in m)

θ is the angle between the incline and the horizontal.

The potential energy of a spring compressed a distance $\vec{s}$ is given by:

$$U_E = \frac{1}{2}k\vec{s}^2$$

where k is the spring constant.

The force exerted by a spring is proportional to the distance the spring is compressed or stretched. That is:

$$\vec{F} = k\vec{s}$$

So the spring constant can be determined by applying different forces to stretch or compress the spring by different amounts. A graph of force versus distance is then a straight line with gradient equal to k.

If energy is conserved, the potential energy in the compressed spring is converted completely into gravitational potential energy.

MATERIALS

- dynamics trolley with plunger
- masses
- pulley with clamp
- string
- mass balance
- dynamics track or board at least 1.5 metres long
- metre ruler
- mass hanger and mass set (several kilograms)

PURPOSE

To examine gravitational potential energy and spring potential energy to show how energy is conserved.

PROCEDURE

Determining the spring constant

1 Level the track by adjusting the levelling feet until a trolley placed on the track does not move. (If there are no levelling feet, use cardboard or books for levelling.)

2 Use the balance to find the mass of the trolley. Record this value.

3 Set the trolley on the track with the spring plunger extended fully and directly against a stopping block. Attach a string to the trolley and attach the other end to a mass hanger, passing the string over the pulley. Record the trolley's position.

4 Add mass to the mass hanger so that the mass is compressing the trolley's plunger and record the new position. Repeat this for a total of 5 different masses. Record each trial in Table 1.

PRACTICAL ACTIVITY 2.6

Gravitational potential energy

1 Remove the string from the trolley and set the spring plunger to its maximum compression position. Place the trolley against the end stop. Measure the distance the spring plunger is compressed and record this value.
2 Incline the track and measure its height and the length of the track to determine the angle of the track. Record the distances measured for checking later and calculate the angle of incline. Record the initial position of the trolley.
3 Release the plunger by tapping it with a stick and record the distance the trolley goes up the track. Repeat this 5 times. Record the maximum distance the trolley went.
4 Change the angle of inclination and repeat the measurements.
5 Add mass to the trolley and repeat the measurements.

DATA AND ANALYSIS

Mass of trolley: __________ kg

Initial distance plunger is compressed __________ m

Percentage difference = $\%\Delta = \frac{U_g - U_E}{U_g} \times \frac{100}{1}$

1 Using your initial data, plot a graph of force versus displacement. Draw a line of best fit through the data points and determine the slope of the line. The slope is equal to the effective spring constant *k*.

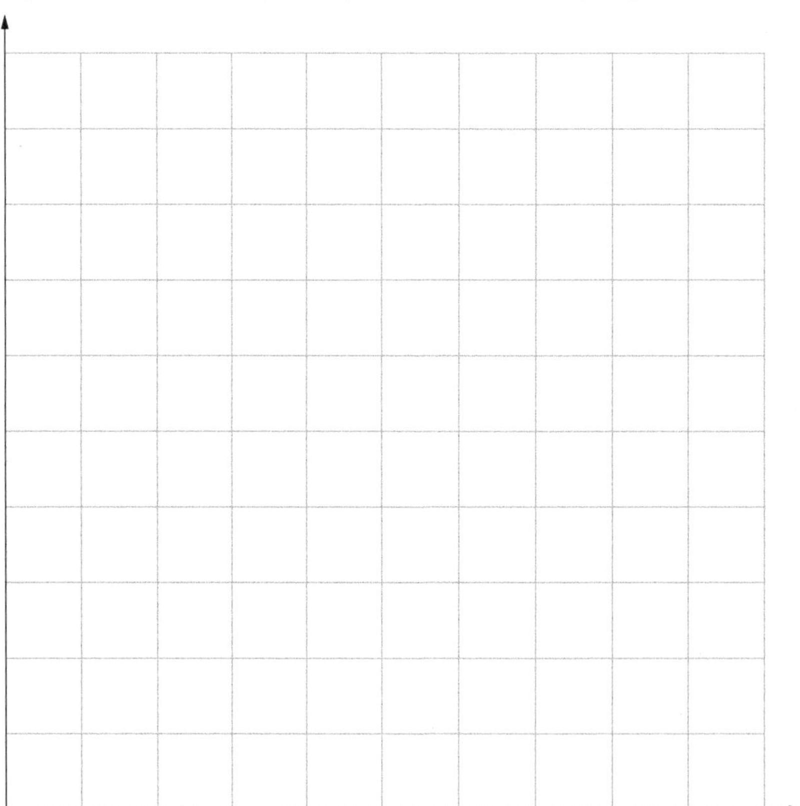

2 Calculate the spring potential energy and record this in Table 1.
3 Calculate the gravitational potential energy for each trial and record this in Table 1.

ISBN 978 1 4886 1935 9

TABLE 1 Spring potential energy

Mass of hanger (kg)	Position (m)	$U_E = \frac{1}{2}k\vec{s}^2$ (J)
0		

4 Calculate the percentage difference between the spring potential energy and the gravitational potential energy and record it in Table 2.

TABLE 2 Gravitational potential energy

Height of track (m)	Track length (m)	θ (°)	Distance travelled (m)	Height travelled $\vec{h} = \vec{s}\sin\theta$(m)	Mass of trolley (kg)	$\Delta U_g = m\vec{g}\Delta\vec{h}$ (J)	% difference

CONCLUSION

1 Comment on your results. Can you state that energy was conserved?

2 Assuming that there was a discrepancy, which of the potential energies was larger? Where did this 'lost' energy go?

3 When the mass of the trolley was doubled, why did the gravitational potential energy remain about the same?

RATING MY LEARNING	My understanding improved	Not confident ◄——► Very confident ○ ○ ○ ○ ○	I answered questions without help	Not confident ◄——► Very confident ○ ○ ○ ○ ○	I corrected my errors without help	Not confident ◄——► Very confident ○ ○ ○ ○ ○

 ISBN 978 1 4886 1935 9

DEPTH STUDY 2.1

Crumple zones and collisions

Suggested duration: 3.5–4 hours

INTRODUCTION

Road accidents are an unfortunate but inevitable part of highway and city driving. Manufacturers take the safety of their customers seriously and invest millions of dollars in research to design a vehicle that prioritises minimising damage to passengers while also seeking to minimise damage to the vehicle. Most modern vehicle designs include sophisticated crumple zones and strategically placed impact absorbers, airbags and more.

This depth study requires you to question how crumple zones work and predict how this can be applied to constructing a working model. You will process and analyse data and information from practical tests and use problem-solving techniques to apply your findings. You will communicate your results in a short individual or group presentation that includes your initial research, design, tests, design modification and final conclusions. Photos of your tests and modelling can be included to illustrate your progress.

MATERIALS

- 1 m^2 cardboard
- 6 sheets of A4 paper
- 6 rubber balloons
- 0.3 m^2 aluminum foil
- 120 g soft clay
- 10 rubber bands
- 20 toothpicks
- 50 cm length of duct tape
- 1 m length of masking tape
- tube of non-epoxy glue
- 10 cm length of Velcro

PURPOSE

Design and construct a crumple zone for a model vehicle that minimises the impact forces experienced in a collision while also minimising size and weight. Your design will be tested in no more than three separate trials, not including initial trials to determine the initial force in your design. The lowest peak force from the three trials will be used to evaluate how well the task was executed.

Design requirements and constraints

- Use only the items listed in the Materials section.
- The design must not impede the trolley's ability to move.
- The design cannot extend more than 5.0 cm in length.
- The design must not add more than 200 g to the mass of the trolley.
- The design must be constructed so that it attaches firmly to the front of the trolley, but can be attached and removed without modifying the trolley itself.
- Three replicate prototypes of your design must be made for testing in three separate trials. (The materials listed should be enough to construct all three prototypes.)

QUESTIONING AND PREDICTING

1 If you were to design a bumper for a model vehicle right now, how would you construct it and what materials, from those listed, would you use? Develop an inquiry question that will be the focus of your design.

2 What variables are being tested?

3 How will you control these variables during testing?

4 Keeping in mind that the bumper should be designed to minimise the force experienced in a collision and cannot be longer than 5.0 cm or have a mass greater than 200 g, sketch two possible designs to explain your proposals, listing three reasons for choosing each design.

5 When a moving object is involved in a collision, its momentum changes. This change in momentum depends on the type of collision that occurs, the time over which it occurs, and whether it is an elastic or inelastic collision. Which type of collision (elastic or inelastic) causes a smaller change in momentum from before to after the collision? Why?

6 When a moving car hits a fixed object, like a wall or telephone pole, what type of collision do you see most often, elastic or inelastic? Explain why you think this type of collision occurs most often.

7 Minimising impulse is important when designing a device to absorb impact. However, a car involved in a collision can experience a large impulse and still not sustain damage. Write a hypothesis explaining how your crumple zone will minimise the impulse in a collision.

 ISBN 978 1 4886 1935 9

CONDUCTING AN INVESTIGATION

8 Using the model vehicle supplied and suitable equipment, run one or more tests with the measuring equipment available to you to determine the peak impact force, before adding the crumple zone. Record your results and calculations.

ANALYSING DATA AND INFORMATION

9 Look back at your original ideas. Based on your testing, would these ideas work? Explain why they work or how your initial thoughts have changed and why.

10 Working in a small group, evaluate other students' designs and agree on the most effective, collaborative design. Draw the final design, including dimensions, materials and material quantities. Record the important design points and your reasons for the design.

Build your prototypes and run the three trials. You may want to take a photo of each prototype or otherwise identify each for reference in your final review. Using the methods available to you, determine the peak force for each prototype.

COMMUNICATING

Communicate your findings with an illustrated group or individual presentation. Your presentation should include your responses to the following questions.

11 What was the biggest reduction in peak force recorded by the class? How did this value compare to your group's lowest peak force value?

 ISBN 978 1 4886 1935 9

12 Look at the designs produced by other groups. What was different about their designs that would have affected their performance relative to yours—what made some more effective and some less effective?

13 What would you change in your design to make it more effective? Why do you think these changes would make the design more effective?

ISBN 978 1 4886 1935 9

Use $\vec{g} = -9.8\,\text{m s}^{-2}$ where required.

Multiple choice

1 Which one or more of these forces is a field force?

A air resistance

B weight

C normal force

D impulse

E magnetism

The following information relates to Questions 2 and 3.

A 1200 kg car travelling at 60 km h^{-1} needs to exert a force of 900 N in order to maintain a constant speed on a straight road.

2 What is the sum of all the resistance forces acting on the car?

A 0 N

B 900 N forwards

C 900 N backwards

D 3300 N forwards

E 3300 N backwards

3 If the driver wants to accelerate at 2.0 m s^{-2} what force must now be applied to overcome resistance?

A 0 N

B 900 N forwards

C 900 N backwards

D 3300 N forwards

E 3300 N backwards

The following information relates to Questions 4–6.

A 5.0 kg box rests on a frictionless horizontal table. It is attached to a 2.0 kg mass by a light, inextensible string via a frictionless pulley, as shown.

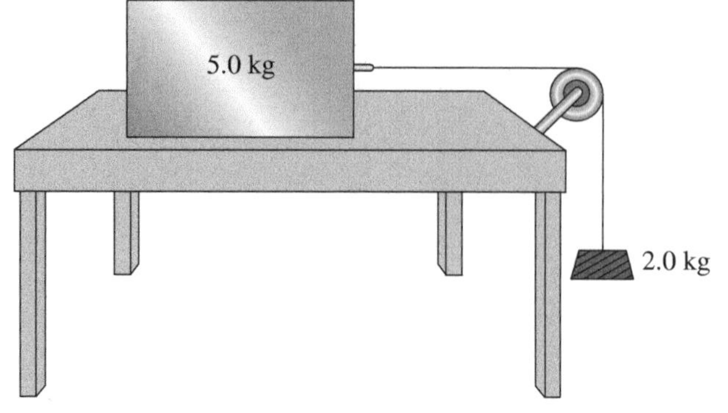

4 What is the acceleration of the box?

A 0 m s^{-2} **B** 2.8 m s^{-2}

C 3.9 m s^{-2} **D** 9.8 m s^{-2}

5 If the table is not smooth, what size frictional force must the table exert to prevent the box from moving?

A 14 N **B** 19.6 N

C 49 N **D** 68.6 N

Short answer

A girl pulls a trolley 2.6 m from rest along a path, using a cord inclined at an angle of 24° to the horizontal. She applies a constant force of 15.0 N. The trolley has a mass of 3.0 kg and the frictional force resisting the motion of the trolley is a constant 9.6 N.

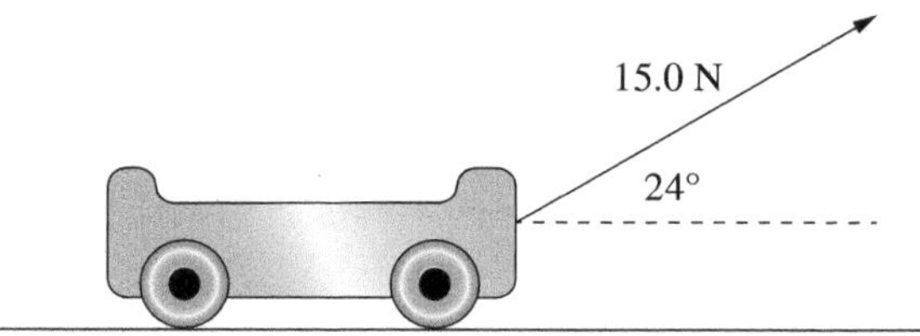

6 Calculate the work done on the trolley by the girl in moving it 2.6 m.

7 Calculate the work done on the trolley by friction.

8 What is the kinetic energy of the trolley after it has moved 2.6 m across the ground?

9 How much work is done on the trolley by the normal contact force? Explain your answer.

 ISBN 978 1 4886 1935 9

Extended response

10 A boy swings off a platform 2.0 m above a river by hanging onto the very end of a rope that is 4.0 m long. The rope is attached to a tree growing on the edge of the river, 5.0 m above the water.

a How fast is the boy going at the lowest point of the swing?

b Several other children now do the same thing, with each child being recorded on video. The recordings are later analysed carefully and it is found that every child, regardless of their mass and when they let go of the rope, hit the water at the same speed. Using your understanding of the conservation of mechanical energy, explain why this occurred.

A train wagon of mass 23 tonnes is rolling along a horizontal track at 8.3 m s^{-1} when it collides with a stationary loaded wagon that has a mass of 34 tonnes. The two wagons become locked together.

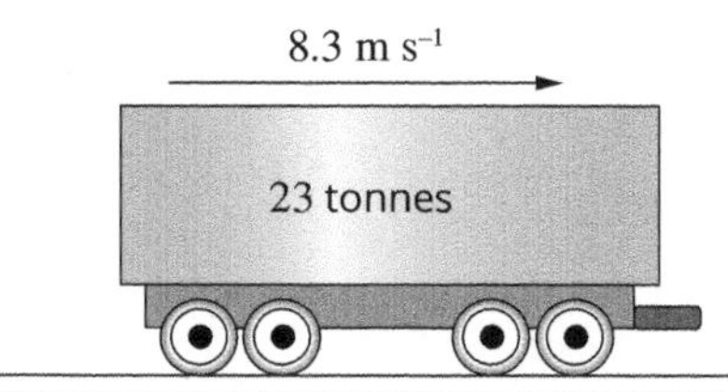

11 a With what speed do the two wagons move off together?

b How much kinetic energy has been lost in the collision?

c Is this an elastic or inelastic collision?

d According to the law of conservation of momentum, when a heavy load is dropped onto the tray of a moving train, the train's mass increases suddenly and its speed is reduced. Would the train then speed up if the load falls off the tray? Explain your answer.

MODULE 3

Waves and thermodynamics

Outcomes

By the end of this module you will be able to:

- conduct investigations to collect valid and reliable primary and secondary data and information PH11-3
- select and process appropriate qualitative and quantitative data and information using a range of appropriate media PH11-4
- solve scientific problems using primary and secondary data, critical thinking skills and scientific processes PH11-6
- communicate scientific understanding using suitable language and terminology for a specific audience or purpose PH11-7
- explain and analyse waves and the transfer of energy by sound, light and thermodynamic principles PH11-10

Content

WAVE PROPERTIES

INQUIRY QUESTION **What are the properties of all waves and wave motion?**

By the end of this module you will be able to:

- conduct a practical investigation involving the creation of mechanical waves in a variety of situations in order to explain: CCT
  - the role of the medium in the propagation of mechanical waves
  - the transfer of energy involved in the propagation of mechanical waves (ACSPH067, ACSPH070)
- conduct practical investigations to explain and analyse the differences between: CCT
  - transverse and longitudinal waves (ACSPH068)
  - mechanical and electromagnetic waves (ACSPH070, ACSPH074)
- construct and/or interpret graphs of displacement as a function of time and as a function of position of transverse and longitudinal waves, and relate the features of those graphs to the following wave characteristics:
  - velocity
  - frequency
  - period
  - wavelength
  - wave number
  - displacement and amplitude (ACSPH069) ICT N
- solve problems and/or make predictions by modelling and applying the following relationships to a variety of situations: ICT N
  - $v = f\lambda$
  - $f = \frac{1}{T}$
  - $k = \frac{2\pi}{\lambda}$

WAVE BEHAVIOUR

INQUIRY QUESTION **How do waves behave?**

By the end of this module you will be able to:

- explain the behaviour of waves in a variety of situations by investigating the phenomena of:
  - reflection
  - refraction
  - diffraction
  - wave superposition (ACSPH071, ACSPH072)
- conduct an investigation to distinguish between progressive and standing waves (ACSPH072)
- conduct an investigation to explore resonance in mechanical systems and the relationships between: CCT
  - driving frequency
  - natural frequency of the oscillating system
  - amplitude of motion
  - transfer/transformation of energy within the system (ACSPH073) ICT N

SOUND WAVES

INQUIRY QUESTION **What evidence suggests that sound is a mechanical wave?**

By the end of this module you will be able to:

- conduct a practical investigation to relate the pitch and loudness of a sound to its wave characteristics
- model the behaviour of sound in air as a longitudinal wave
- relate the displacement of air molecules to variations in pressure (ACSPH070)
- investigate quantitatively the relationship between distance and intensity of sound
- conduct investigations to analyse the reflection, diffraction, resonance and superposition of sound waves (ACSPH071)
- investigate and model the behaviour of standing waves on strings and/or in pipes to relate quantitatively the fundamental and harmonic frequencies of the waves that are produced to the physical characteristics (eg length, mass, tension, wave velocity) of the medium (ACSPH072) ICT N
- analyse qualitatively and quantitatively the relationships of the wave nature of sound to explain: CCT
  - beats ($f_{beat} = |f_2 - f_1|$)
  - the Doppler effect $f' = f\frac{(v_{wave} + v_{observer})}{(v_{wave} - v_{source})}$

Module 3 • Waves and thermodynamics

RAY MODEL OF LIGHT

INQUIRY QUESTION **What properties can be demonstrated when using the ray model of light?**

By the end of this module you will be able to:

- conduct a practical investigation to analyse the formation of images in mirrors and lenses via reflection and refraction using the ray model of light (ACSPH075)
- conduct investigations to examine qualitatively and quantitatively the refraction and total internal reflection of light (ACSPH075, ACSPH076)
- predict quantitatively, using Snell's Law, the refraction and total internal reflection of light in a variety of situations CCT
- conduct a practical investigation to demonstrate and explain the phenomenon of the dispersion of light CCT
- conduct an investigation to demonstrate the relationship between inverse square law, the intensity of light and the transfer of energy (ACSPH077)
- solve problems or make quantitative predictions in a variety of situations by applying the following relationships to: ICT N
  - $n_x = \frac{c}{v_x}$ – for the refractive index of medium x, v_x is the speed of light in the medium
  - $n_1 \sin i = n_2 \sin r$ (Snell's law)
  - $\sin i_c = \frac{1}{n_x}$ – for the critical angle i_c of medium x
  - $I_1 r_1^2 = I_2 r_2^2$ – to compare the intensity of light at two points, r_1 and r_2

THERMODYNAMICS

INQUIRY QUESTION **How are temperature, thermal energy and particle motion related?** CCT

By the end of this module you will be able to:

- explain the relationship between the temperature of an object and the kinetic energy of the particles within it (ACSPH018)
- explain the concept of thermal equilibrium (ACSPH022)
- analyse the relationship between the change in temperature of an object and its specific heat capacity using the equation $\Delta Q = mc\Delta T$ (ACSPH020)
- investigate energy transfer by the process of:
  - conduction
  - convection
  - radiation (ACSPH016)
- conduct an investigation to analyse qualitatively and quantitatively the latent heat involved in a change of state
- model and predict quantitatively energy transfer from hot objects by the process of thermal conductivity CCT
- apply the following relationships to solve problems and make quantitative predictions in a variety of situations: ICT N
  - $\Delta Q = mc\Delta T$, where c is the specific heat capacity of a substance
  - $\frac{Q}{t} = \frac{kA\Delta T}{d}$, where k is the thermal conductivity of a material

Key knowledge

Wave properties

Waves are everywhere, and they come in many shapes and forms. Vibrating objects transfer energy through waves, travelling outwards from the source.

MECHANICAL WAVES

Waves are classified by what they move through. **Electromagnetic waves**, or light waves, are able to travel without the need for a medium. For example, light from the Sun is able to reach the Earth through the vacuum of space. A **mechanical wave** is any wave where the energy is transferred through vibrations in a medium. Waves on water, in a string and in air are examples of mechanical waves.

A wave may be a single pulse, or it may be continuous or periodic (successive crests and troughs or compressions and rarefactions). A wave only transfers energy from one point to another. There is no transfer of matter, although matter may move as the wave passes through it.

Mechanical waves can be either transverse or longitudinal. In a **transverse** wave, the oscillations are perpendicular to the direction in which the wave energy is travelling, so the particles in the medium move sideways (transversely) as the waves pass (Figure 3.1a). In a **longitudinal** wave, the oscillations are parallel to (along) the direction the wave energy is travelling, so the particles in the medium move back and forth (longitudinally) as the wave passes (Figure 3.1b). Sound is an example of a longitudinal wave.

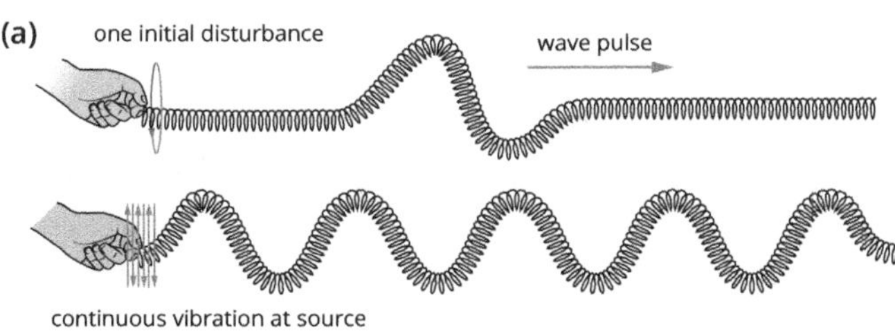

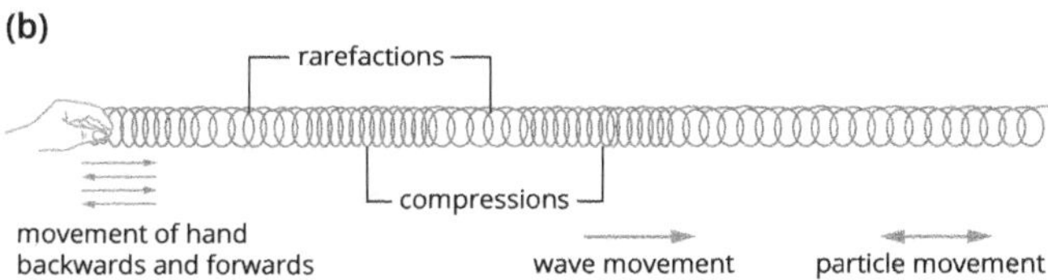

FIGURE 3.1 (a) A transverse wave, and (b) a longitudinal wave

Measuring mechanical waves

Waves can be represented by displacement–distance graphs and displacement–time graphs. From a displacement-–distance graph, the amplitude and wavelength can be easily determined.

The **amplitude** of a wave is the maximum displacement of the particles from the rest position (when displacement is zero).

The **wavelength**, denoted by the Greek letter λ is measured in metres and is equal to the distance between any two repeating points, also called a cycle. An example of a displacement–distance graph, illustrating both amplitude and wavelength, is shown in Figure 3.2.

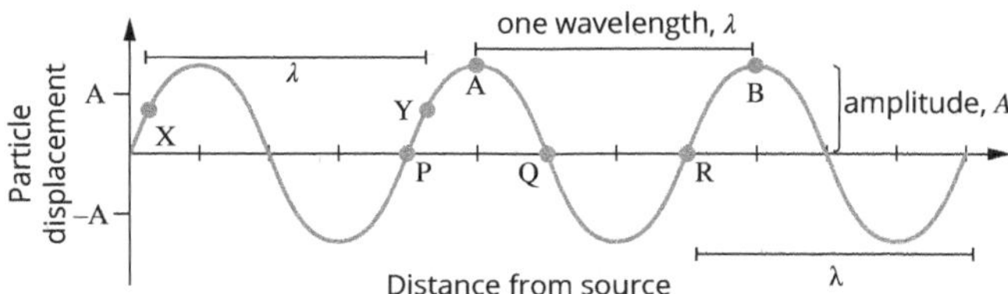

FIGURE 3.2 Displacement–distance graph

The **wavenumber**, k, is the number of waves per unit distance. It has an inverse relationship to the wavelength:

$$k = \frac{2\pi}{\lambda}$$

The **period** is the time it takes for any point on the wave to go through one complete cycle and the **frequency** is the number of complete cycles that pass a given point per second. The period of a wave has an inverse relationship to the frequency, according to the relationship:

$$T = \frac{1}{f}$$

The rate at which energy can travel can also be determined, if either frequency (or period) and wavelength is provided. The speed of a wave can be calculated using the wave equation:

$$v = f\lambda = \frac{\lambda}{T}$$

Amplitude, wavelength, wavenumber, frequency and period can all be determined from a displacement–time graph. The amplitude and period can be read straight from the graph (Figure 3.3).

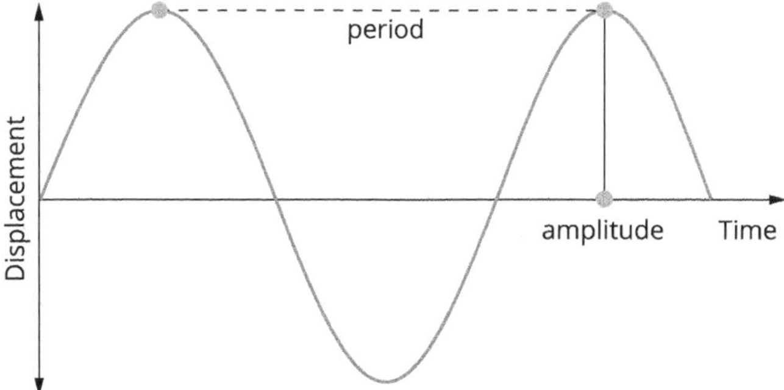

FIGURE 3.3 Reading amplitude and period from a displacement–distance graph

Wave behaviour

All waves exhibit certain behaviours when reaching a boundary. They can undergo refraction, reflection, interference and diffraction.

WAVE INTERACTIONS

A wave that meets a boundary between two media can be partly reflected, partly transmitted and partly absorbed. This interaction can be thought of as involving three different waves:

- the incident wave, which is the wave as it approaches the boundary
- the transmitted wave, which is the wave that continues into the second medium
- the reflected wave, which is the wave that bounces back into the first medium.

Reflection

When a wave meets a boundary, the way it is reflected depends on the nature of the boundary:

- A wave reflecting from a fixed boundary is illustrated in Figure 3.4a, where the rope is fixed to the post. In this case the wave undergoes a phase change of 180° (π radians). That is, crests reflect as troughs and troughs reflect as crests.
- A wave reflecting from a free boundary is illustrated in Figure 3.4b, where the rope is not fixed to the post and is free to move up and down. In this case the wave does not undergo a phase change. That is, crests reflect as crests and troughs reflect as troughs.

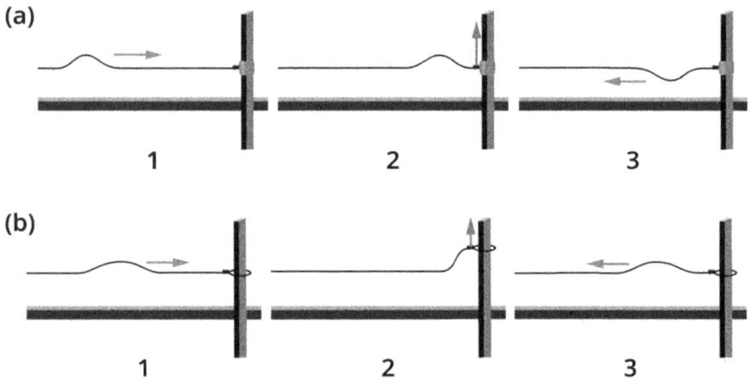

FIGURE 3.4 (a) A phase change at a fixed boundary, and (b) no phase change at a free boundary

Another example of this concept is a lighter rope that is connected to a heavier rope, such as in Figure 3.5a. When the incident wave meets the heavy section, it undergoes a phase change, because the heavy rope acts like a fixed boundary. Figure 3.5b illustrates the reflected wave travelling back into the light rope, and the transmitted wave continuing forwards into the heavy rope. When the wave is transmitted into the heavy rope, a portion of the energy of the wave is absorbed by the boundary, so it has less energy. The amplitude of the transmitted wave is therefore smaller than the amplitude of the incident incident wave.

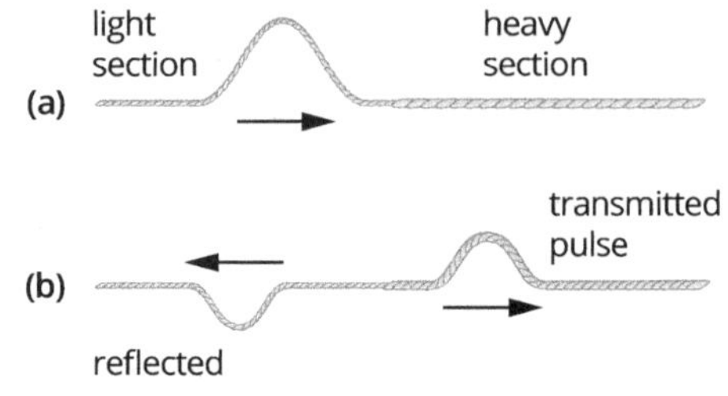

FIGURE 3.5 Change in medium of rope

When a wave (or part of a wave) is reflected from a boundary, the angle of reflection equals the angle of incidence (Figure 3.6). This is known as the law of reflection. The angles are measured from the the normal, which is an imaginary line at right angles to the boundary at the point where the wave meets it.

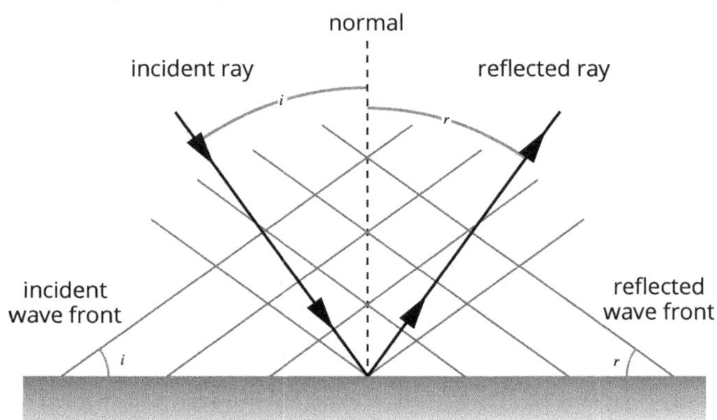

FIGURE 3.6 The law of reflection

 If a wavefront meets an irregular or rough surface, the resulting reflection is spread over a wide range of angles. This is known as diffuse reflection.

When waves pass from one medium into a different medium, other changes in the wave can be observed. The frequency does not change, but the speed and wavelength change, and the direction of propagation of the wave can also change.

Refraction

Refraction is a change in the direction of propagation of a wave that occurs when it passes from one medium into another, as shown in Figure 3.7. The lines in this figure represent wavefronts, which can be thought of as the crests of consecutive waves. As the wavefronts pass from the glass into the air, they bend farther away from the normal.

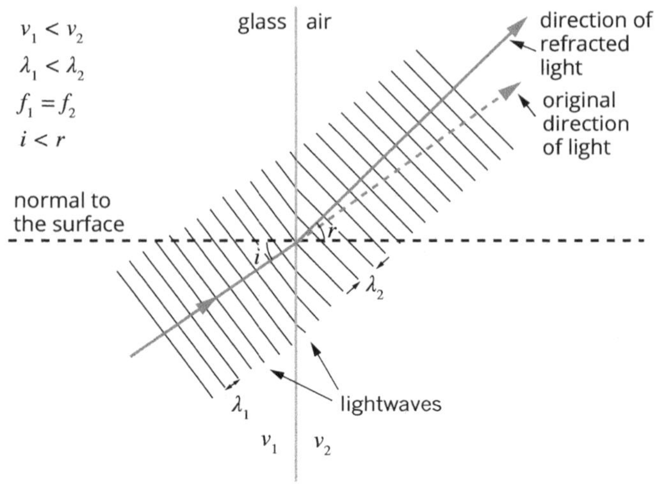

FIGURE 3.7 Refraction

ISBN 978 1 4886 1935 9

The angle of incidence *i*, which is defined as the angle between the direction of propagation and the normal, is less than the angle of refraction *r*.

Diffraction

When a plane (straight) wave passes through a narrow opening or meets a sharp object, it experiences diffraction. **Diffraction** is the bending of the wavefronts as they pass through the opening. The diffraction is large if the wavelength is similar to, or larger than, the width of the opening, as shown in Figure 3.8a. When the wavelength is smaller than the width of the opening, the diffraction is less (Figure 3.8b).

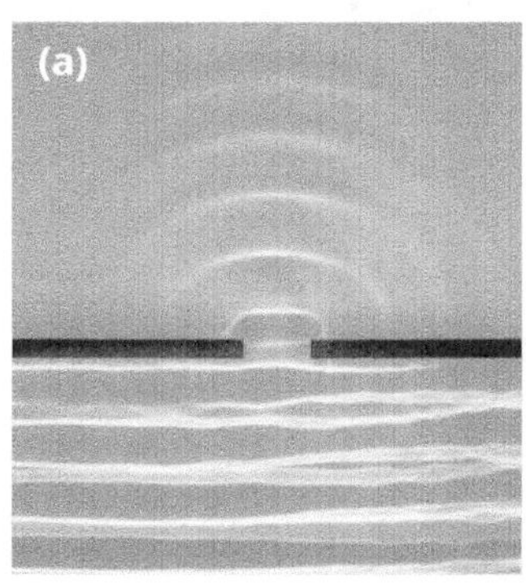

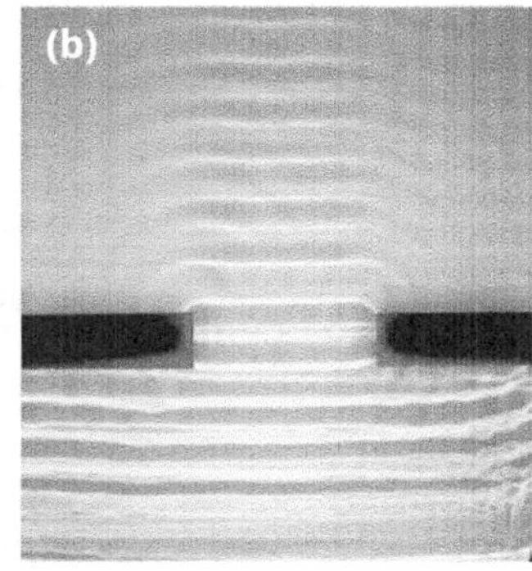

FIGURE 3.8 (a) When the wavelength is similar to or larger than the width of the opening, the diffraction is large. (b) When the wavelength is smaller than the width of the opening, the diffraction is less.

When two waves meet, they can interact by:

- producing a smaller wave (if the crests of one wave meets the troughs of the other wave); this is referred to as destructive interference
- producing a larger wave or supercrest (if the crests of each wave meet); this is known as constructive interference.

This interference can occur in both mechanical and electromagnetic waves, and is called **superposition**. The principle of superposition states that when two or more waves interact, the resultant displacement or pressure at each point along the wave is the vector sum of the displacements or pressures of the component waves. The effect of superposition is shown in Figure 3.9, where the waves interact to produce a larger wave (Figure 3.9b).

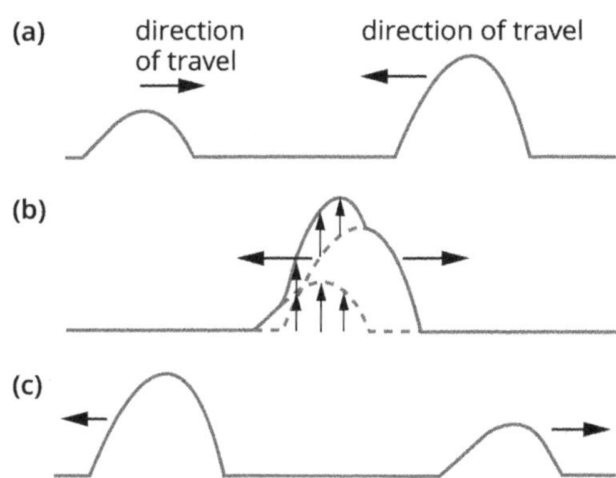

FIGURE 3.9 Superposition of two waves

Resonance

An object that can vibrate tends to do so at a specific frequency known as the natural or resonant frequency. **Resonance** occurs when a source produces a driving frequency (also called a forcing vibration) equal to the natural frequency of the object. Two special effects occur with resonance:

- the amplitude of the vibration increases
- the maximum possible energy from the source is transferred to the resonating object.

A child's swing in a playground is a good example of a resonant system. When the swing is pushed once, the movement of the swing represents an object vibrating at its natural frequency. If the swing continues to be pushed at this natural frequency, the amplitude of the swing increases. However, if the frequency of the pushes does not match the natural frequency of the swing, pushing becomes more difficult and the amplitude will not increase.

Sound waves

Sound waves transfer energy through the vibration of molecules. For this to happen, sound must travel through a medium. Sound is a longitudinal wave, so the oscillations are in the direction that the wave is travelling.

Consider the loudspeaker shown in Figure 3.10a, which has a cone that vibrates in and out. When the cone pushes outwards, the molecules in the air are compressed, producing a region of high pressure (a **compression**). When the cone then moves inwards, a region of lower pressure (a **rarefaction**) is produced. This compression and rarefaction forms a wave that propagates away from the loudspeaker. As the loudspeaker cone continues to vibrate, a regular series of compressions and rarefactions is produced.

In Figure 3.10a, Points W and X are in consecutive compressions and are said to be in phase, which means they are one wavelength apart. Conversely, points Y and Z are one wavelength apart as they are located at adjacent rarefactions.

This information can be modelled using a pressure versus distance graph. This graph is represented like a transverse wave shown in Figure 3.10b. The compressions and rarefactions of a longitudinal wave coincide with the peaks and troughs respectively of the transverse wave representation.

Because sound is a wave, it behaves like any other wave. Echoes are sound waves reflecting off surfaces. The controlled reflection of sound waves allows images to be created, providing many uses in medicine and industry. For example, sonar (sound navigation and ranging) uses the reflecting properties of sound waves in water for navigating in shallow waters, detecting submerged objects such as shipwrecks, and finding schools of fish.

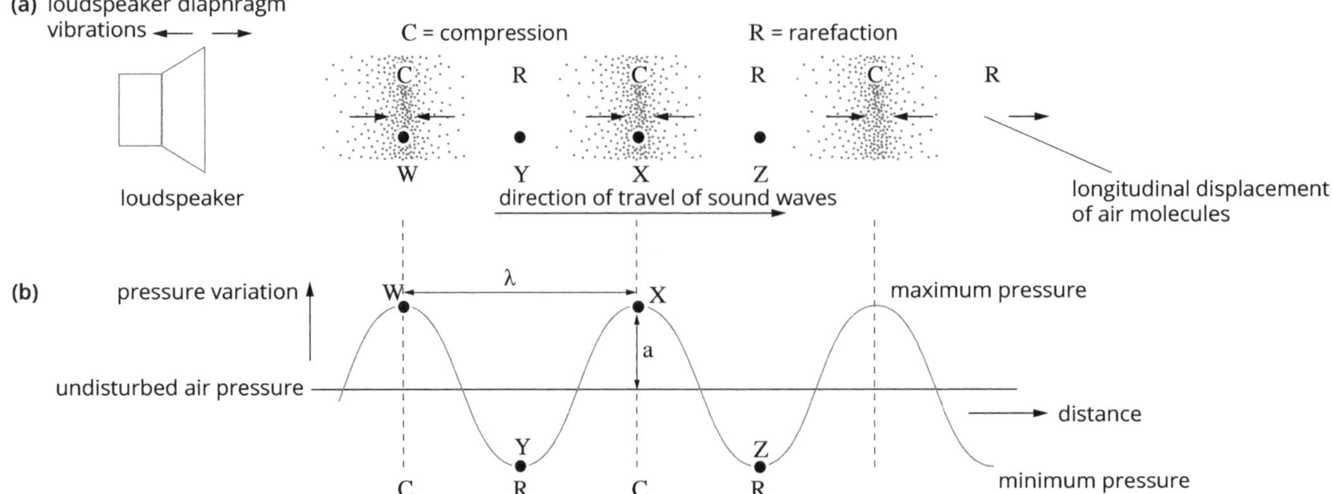

FIGURE 3.10 (a) A longitudinal sound wave of compressions and rarefactions produced by a vibrating loudspeaker and (b) a graphical representation of the same wave, with pressure plotted against distance from the loudspeaker

SOUND BEHAVIOUR

Sound intensity is measured by the wave's amplitude and frequency. A higher intensity, or a larger amplitude, is equivalent to an increase in volume. Sounds that have a higher pitch have a greater frequency. The **intensity** (I) of a sound wave describes the rate at which sound energy passes through an area; it is measured in watts per square metre ($W\,m^{-2}$). For example, a sound intensity of $10^{-12}\,W\,m^{-2}$ is the lowest level of sound heard by people. In contrast, a sound intensity of $1\,W\,m^{-2}$ would cause physical pain to a human.

The intensity of sound follows an inverse square relationship: $I \propto \frac{1}{r^2}$, where r = distance from the source (in m). Using this relationship, it is possible to calculate the intensity of sound. For example, the sound from a fire alarm might be $1.0\,mW\,m^{-2}$ at a distance of 2.0 m. The intensity of the sound at a distance of 5 m would then be:

$$\frac{I_1}{I_2} \propto \frac{r_2^2}{r_1^2}$$

$$I_2 = \frac{I_1 r_1^2}{r_2^2}$$

$$= \frac{1 \times 2^2}{5^2}$$

$$= 0.16\,mW\,m^{-2}$$

The Doppler effect is a phenomenon that is observed whenever there is relative movement between the source of waves and an observer. It causes an apparent increase in frequency when the relative movement is towards the observer, and an apparent decrease in frequency when the relative movement is away from the observer.

The most common example of this phenomenon is the change of pitch heard when a vehicle such as an ambulance is approaching or receding, as in Figure 3.11.

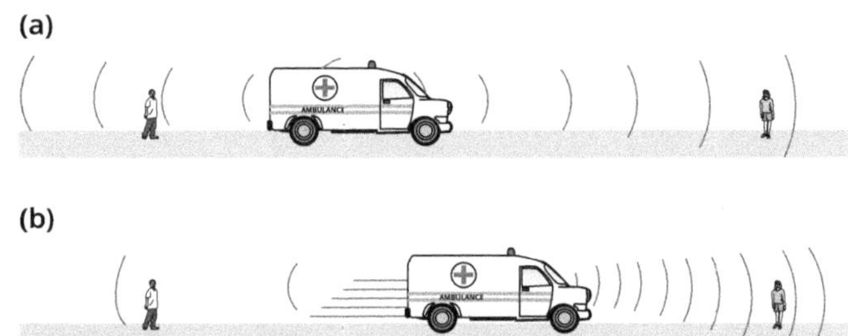

FIGURE 3.11 The Doppler effect. (a) The ambulance is stationary, so the frequency heard by the two observers is the same. (b) The ambulance is moving to the right, so the observer in front of the ambulance hears a higher frequency and the observer behind the ambulance hears a lower frequency.

For a mechanical wave, the Doppler effect can be caused by the relative motion of the source and the observer, and also by the motion of the medium (e.g. wind). For waves that are not mechanical, such as light and other electromagnetic waves, only the relative motion of observer and the source can cause the Doppler effect.

The altered frequency, f', can be calculated using the following formula:

$$f' = f\frac{(v_{wave} + v_{observer})}{(v_{wave} - v_{source})}$$

where:

f' is the apparent or observed frequency (in Hz)

f is the original frequency (in Hz)

v_{wave} is the speed of the waves in the medium (in $m\,s^{-1}$)

$v_{observer}$ is the speed of the observer relative to the medium (in $m\,s^{-1}$)

v_{source} is the speed of the source relative to the medium (in $m\,s^{-1}$).

 ISBN 978 1 4886 1935 9

> ℹ $v_{observer}$ is positive if the observer is moving towards the source, and negative if moving away.
>
> v_{source} is positive if the source is moving towards the observer, and negative if moving away.

When two sound waves are present, superposition leads to interference that is either constructive or destructive. When two waves with equal amplitudes but different frequencies interact, a pulse or beat with a rising and falling amplitude is created. This is shown in Figure 3.12.

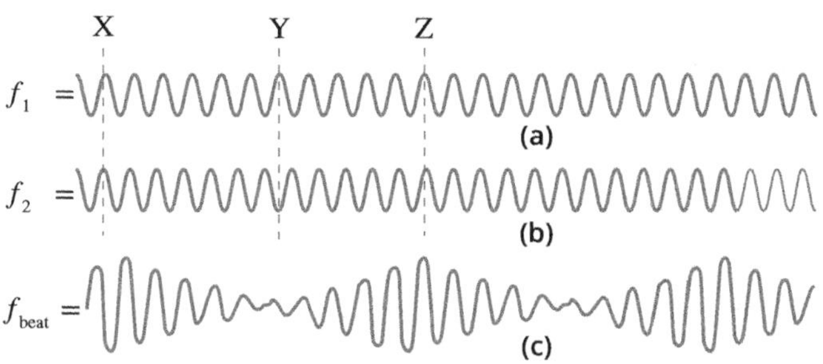

FIGURE 3.12 A pulse or beat with a rising and falling amplitude is created when two waves with equal amplitudes interact. The beat wavelength is the distance between X and Z. A trough occurs at point Y, where the amplitudes cancel each other out.

The **beat frequency** of this pulse can be found with the following formula:

> $f_{beat} = |f_1 - f_2|$
>
> where f_1 and f_2 are the two sound frequencies and $|f_1 - f_2|$ is the absolute or positive value of the difference.

This phenomenon can be used to tune musical instruments. For example, if two violins are not in tune, continuously playing the same note will produce a beat. By adjusting the string tension on one violin, the beats will get farther and farther apart, until eventually the two instruments will be in tune.

STANDING WAVES

Standing or stationary waves occur as a result of resonance at the natural frequency of vibration. They occur when two waves of the same amplitude and frequency are travelling in opposite directions in the same string, or in a pipe.

Nodes are produced by destructive interference, in which the amplitudes between the two waves completely cancel.

Antinodes represent constructive interference, where the amplitudes from the two waves combine.

Standing waves can be produced by vibrating a string attached to a fixed point, as shown in Figure 3.13. Standing waves with different can be produced by vibrating the string at different frequencies.

Standing sound wave frequencies are referred to as **harmonics**. The simplest mode is called the fundamental frequency or first harmonic. The fundamental frequency has the longest wavelength and the lowest frequency. The fundamental frequency of an instrument depends on the type of instrument and whether the ends are fixed or not.

Harmonics in strings

For a string fixed at both ends, the wavelength of the standing waves corresponding to the various harmonics is given by: $\lambda = \frac{2l}{n}$

where:

λ = wavelength (in m)

l = length of string (in m)

n = is the number of the harmonic.

The first four harmonics in a string are shown in Figure 3.14. Note that there is a node is at each fixed end. Using the wave equation ($v = f\lambda$) it is possible to find the relationship between frequency, velocity and string length. The general formula is:

$$f = \frac{nv}{2l}$$

All harmonics may be present.

For a string fixed at one end, the wavelength of the standing waves corresponding to the various harmonics is:

$$\lambda = \frac{4l}{n}$$

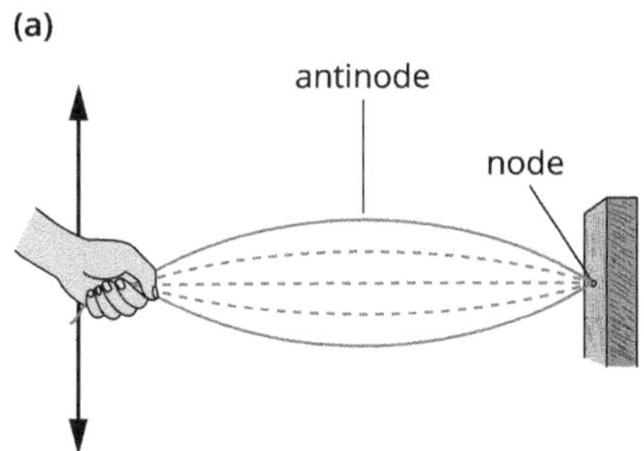

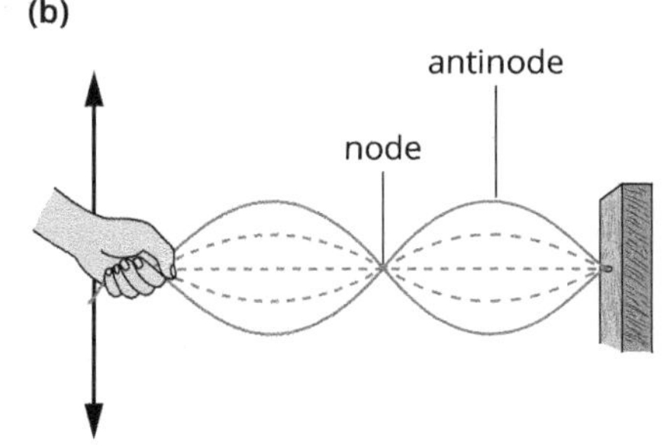

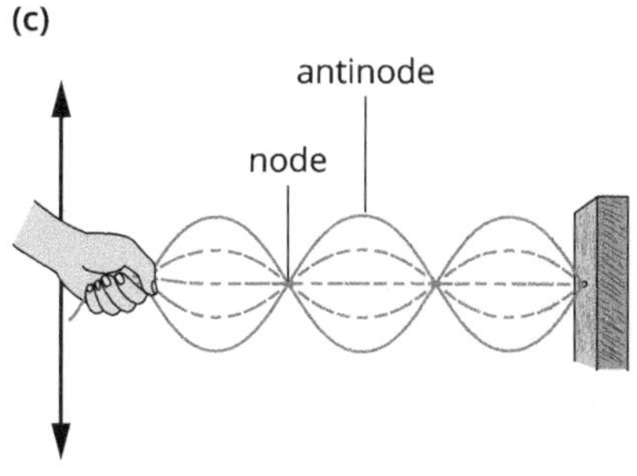

FIGURE 3.13 Standing waves produced by vibrating a string attached to a fixed point, showing vibrations at (a) the fundamental frequency, (b) twice the fundamental frequency, and (c) three times the fundamental frequency

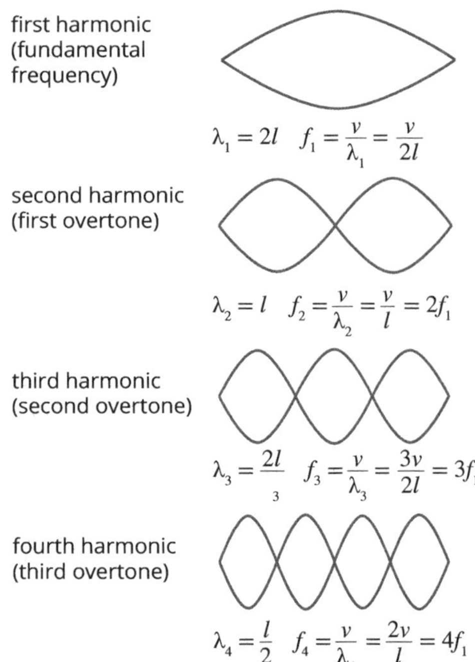

FIGURE 3.14 Harmonics in a string with fixed ends

The first three resonant frequencies for a string fixed at one end are shown in Figure 3.15. There is always a node at the fixed end and an antinode at the free end. As a result, only odd-numbered harmonics can be produced.

The frequency is given by:

$$f = \frac{nv}{4l}$$

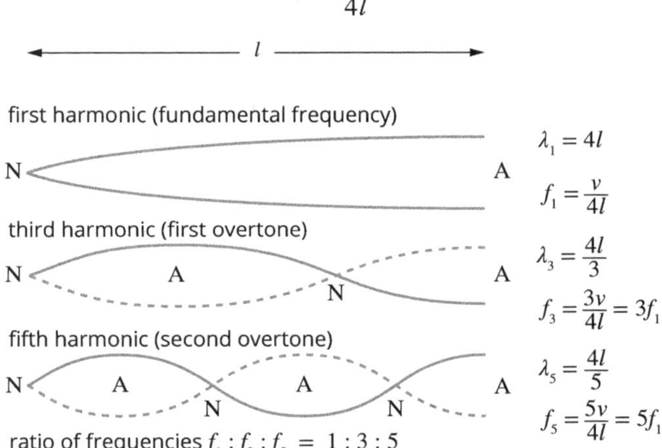

FIGURE 3.15 Harmonics in a string with one free end

Harmonics in air columns or pipes

Many musical instruments consist of air columns or pipes. They include all wind instruments such as flutes, oboes and didgeridoos, all brass instruments such as trumpets and trombones, and some percussion instruments such as tubular bells and chimes.

Pipes open at both ends have antinodes at both ends, whereas pipes with one closed end have a node at one end and an antinode at the other end.

For a tube that is open at both ends, the wavelength of the standing waves corresponding to the various harmonics is:

$$\lambda = \frac{2l}{n}$$

The first four harmonics for a pipe that is open at both ends are the same as those shown in Figure 3.14 for a string with fixed ends. In general for such pipes:

$$f = \frac{nv}{2l}$$

For a tube that is closed at one end, the wavelength of the standing waves corresponding to the various harmonics is:

$$\lambda = \frac{4l}{n}$$

There will be an antinode at the open end and a node at the closed end. This means only odd-numbered harmonics can exist, as shown in Figure 3.16. The frequency is:

$$f = \frac{nv}{4l}$$

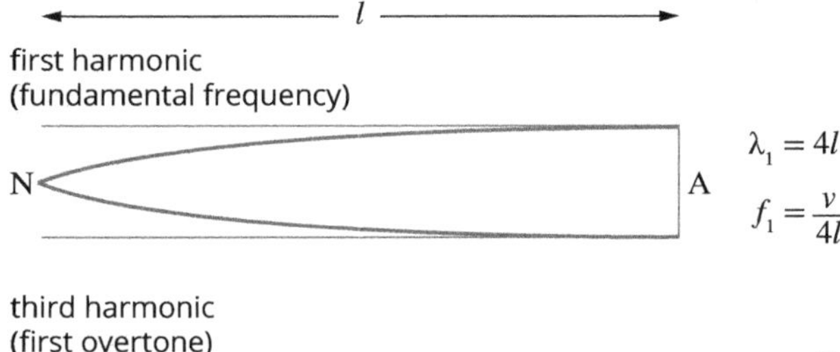

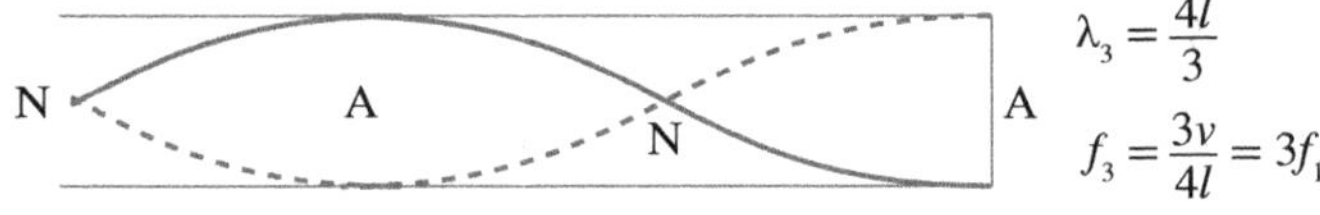

fifth harmonic (second overtone)

N A N A N A

$\lambda_5 = \frac{4l}{5}$

$f_5 = \frac{5v}{4l} = 5f_1$

ratio of frequencies $f_1 : f_3 : f_5 = 1 : 3 : 5$

FIGURE 3.16 Harmonics for an air column with one closed end

Effects of string tension and mass

From these equations it is possible to see how the harmonic frequencies are affected by the length of the pipe or string and the speed of the wave. For string instruments, two other variables affect the frequency: the tension and the mass of the string.

The speed of a wave on a string can be calculated with the following equation:

$$v = \sqrt{\frac{T}{m/l}}$$

where:

T is the tension in the string (in N)

m is the mass (in kg)

l is the length (in m).

 ISBN 978 1 4886 1935 9

This can be included in the equations for frequency for a string fixed at both ends:

$$f = \frac{nv}{2l} = \frac{n}{2l}\sqrt{\frac{T}{m/l}}$$

Frequency is therefore proportional to the square root of tension, so a string with a greater amount of tension will produce a higher frequency.

Similarly, frequency is inversely proportional to the mass of the string. A heavier string will vibrate at a lower frequency if the length and tension are the same.

Ray model of light

Light exhibits behaviours that are characteristic of both waves and particles. For this reason, light can be modelled as either a wave or particles. One very useful model of light as a wave is the ray model of light.

LIGHT AS A RAY

The ray model is not an exact model of light: it is a simple representation of light acting like a wave. It is obtained by choosing a line that is perpendicular to the wavefronts, pointing in the direction of propagation. Because light has the characteristics of a wave, it behaves like any wave. It can undergo refraction, reflection, interference and diffraction.

When light strikes a boundary between two media, some of the light is absorbed and some may be transmitted across the boundary (if the second medium is transparent or translucent). The rest of the incident light is reflected. The law of reflection states that the angle of incidence i is equal to the angle of reflection r.

Only regular reflection can produce a distinct image. Diffuse reflection produces an indistinct image, or no image at all.

Figure 3.17 represents regular reflection in a plane mirror. Ray diagrams can be used to explain the images seen in plane mirrors. These light rays are represented by the straight lines, with arrows indicating the direction in which the light is travelling.

The image formed by a plane mirror is always a virtual image. A **virtual image** means that the light rays only appear to come from the image; they do not actually come from it.

A **real image** is an image that light rays actually come from.

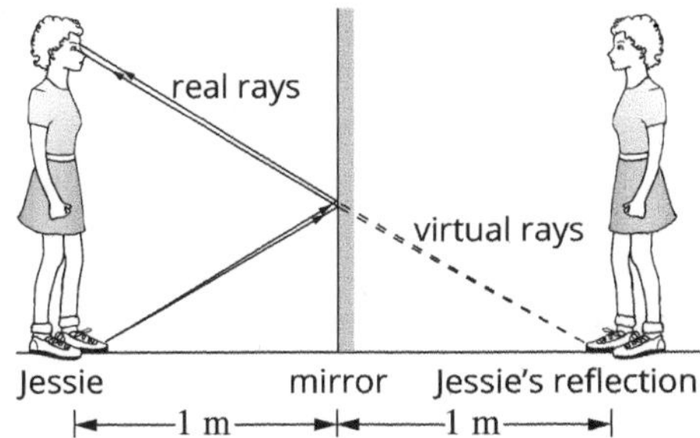

FIGURE 3.17 Regular reflection from a plane mirror always forms a virtual image.

Luminous intensity is a measure of brightness or the rate at which light energy passes through an area of space.

The SI unit for luminous intensity is the candela (cd). The relationship between the intensity of light at two points at different distances from a light source is given by:

$$I_1 r_1^2 = I_2 r_2^2$$

where:

I_1 and I_2 intensities at points 1 and 2 respectively (in cd)

r_1 and r_2 are distances of points 1 and 2 respectively from the light source (in m).

A similar approach to that for the intensity of sound can be taken here. For example, if a flashlight has a light intensity of 20 cd at a distance of 1.5 m from the lens, the intensity of the flashlight 20 m from the lens will be:

$$\frac{I_1}{I_2} \propto \frac{r_2^2}{r_1^2}$$

$$I_2 = \frac{I_1 r_1^2}{r_2^2} = \frac{20 \times 1.5^2}{20^2} = 0.11\ \text{cd}$$

REFRACTION

Refraction is a change in the direction of light when it moves from one medium to another. Refraction is caused by changes in the speed of light. A convenient way to describe the change in speed of a wave is a property called the **refractive index**. The refractive index determines how far light is refracted when entering a material.

The refractive index, n_x, of material x is given by the following formula:

$$n_x = \frac{c}{v_x}$$

where:

c is the the speed of light in a vacuum ($3 \times 10^8\ \text{m s}^{-1}$)

v_x is the the speed of light in material x.

The term 'optical density' refers to the ability of light to pass through a medium, and it is related directly to the concept of refractive index. If a medium has a higher refractive index, it is more optically dense.

When light moves from one medium to another, the changes in speed can be calculated using the following formula:

$$n_1 v_1 = n_2 v_2$$

where:

n_1 the refractive index of the first medium

v_1 speed of light in the first material (in m s^{-1})

n_2 the refractive index of the second medium

v_2 is the speed of light in the second medium (in m s^{-1}).

This relationship can be linked to Snell's law:

$$n_1 \sin i = n_2 \sin r$$

Snell's law relates to the geometry of the situation when a light ray refracts as it moves from one medium to another.

When waves travel from a more optically dense medium to a less optically dense medium, total internal reflection can occur. **Total internal reflection** occurs when the incident angle is equal to or greater than what is known as the critical angle, which can be calculated by the following formula:

$$\sin i_c = \frac{1}{n_x}$$

where i_c is the the critical angle.

White light consists of a range of different frequencies, which we see as different colours. These colours travel at different speeds, so each colour is refracted by a different amount when it passes from one medium into another. This causes white light to split up into different colours, which we call a spectrum.

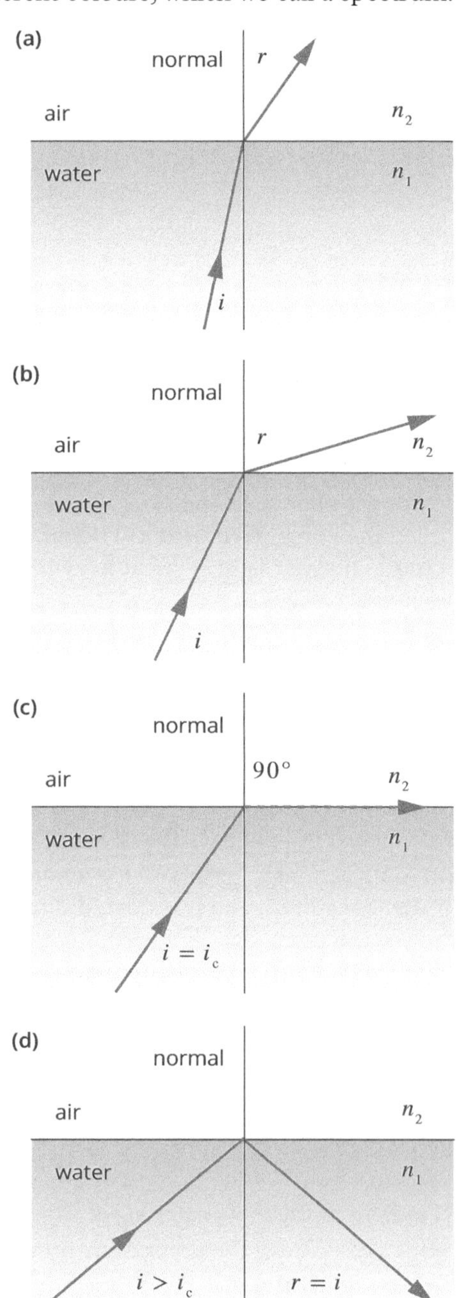

FIGURE 3.18 Internal reflection at an air–water boundary

CURVED MIRRORS AND LENSES

So far we have only considered how light behaves when it meets a flat surface. But when the reflecting or refracting surface is curved, the situation is much more complex.

Curved mirrors and lenses can be classified as either converging or diverging. Converging means that the light rays are reflected or refracted so that they meet at a point, called the focal point. The focal point of a mirror or lens is the point where parallel rays are focused. Concave mirrors are converging mirrors (Figure 3.19a). Converging lenses have one or both surfaces convex.

Diverging means that the light rays spread out, and the rays can be traced backwards until they intersect at a point. Diverging mirrors are convex (Figure 3.19b). Diverging lenses have at least one surface concave.

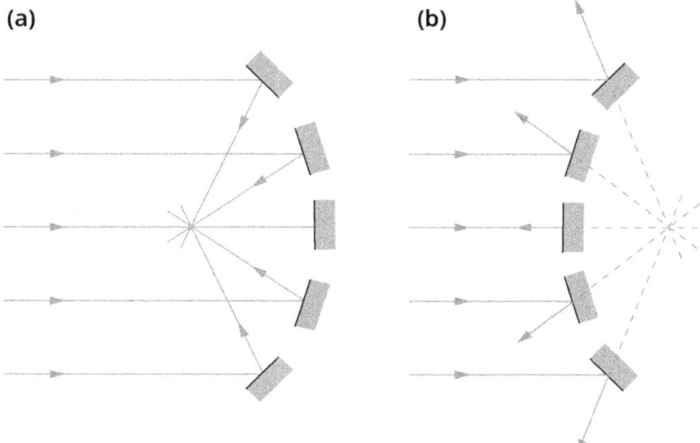

FIGURE 3.19 (a) A concave mirror, showing converging rays forming a real image. (b) A convex mirror, showing converging rays forming a virtual image.

The distance from the mirror to the focal point is called the focal length, and is related to the radius of curvature of the mirror. A convex mirror always forms a virtual and upright image behind the mirror, no matter where the object is situated in front of the mirror. A concave mirror can form either a virtual or a real image, depending on where the object is situated.

Figures 3.20 illustrates images formed by a concave mirror, and Figure 3.21 illustrates images formed by a convex mirror. In these figures, F is the focal point and C is the centre of curvature of the mirror.

The position of the image of an object viewed using a curved mirror or lens can be determined using the mirror/lens formula. This formula assumes that the focal length of a concave mirror or convex lens is a positive value and the focal length of a convex mirror or concave lens is negative. A negative answer for the image distance means that the image is virtual.

The size of an image is often expressed in terms of magnification. Magnification is the ratio of the size of the image to the size of the object. It can be calculated using the following formula.

 ISBN 978 1 4886 1935 9

Mirror/lens formula

$$\frac{1}{f}=\frac{1}{u}+\frac{1}{v}$$

where:

f is the the focal length of the mirror

u is the the distance between the object and the mirror

v is the the distance between the image and the mirror.

The units for distance must be the same.

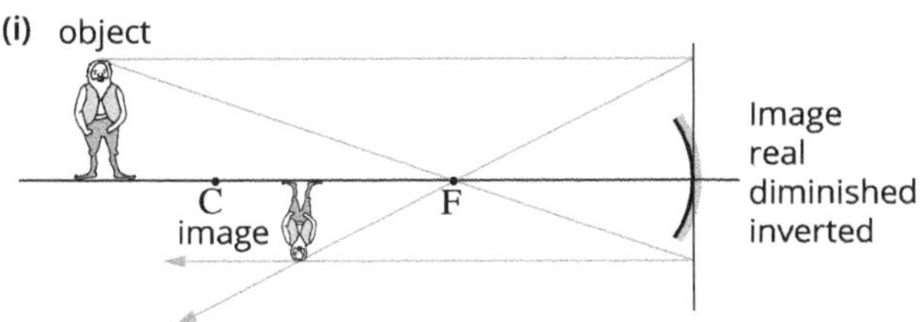

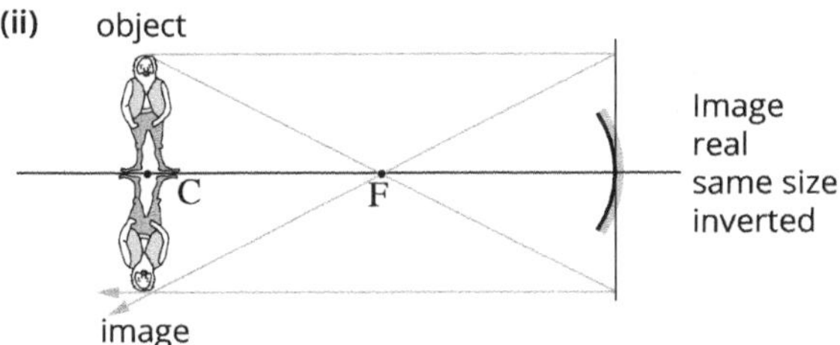

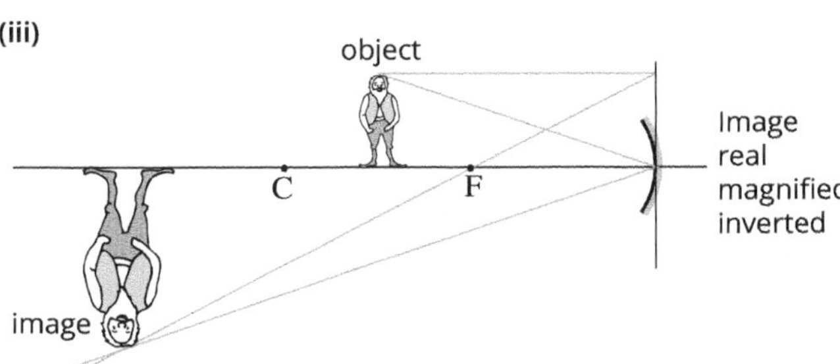

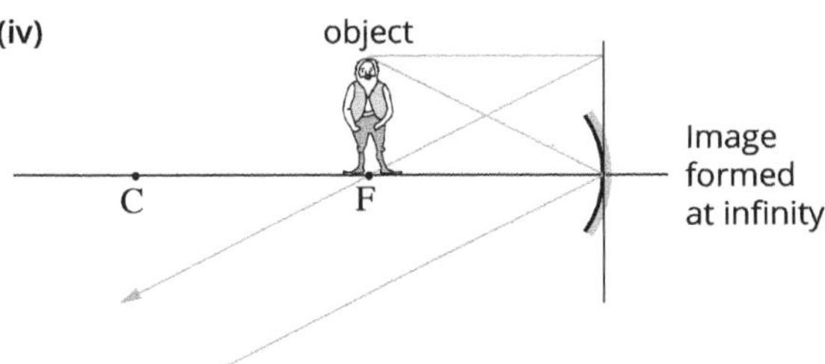

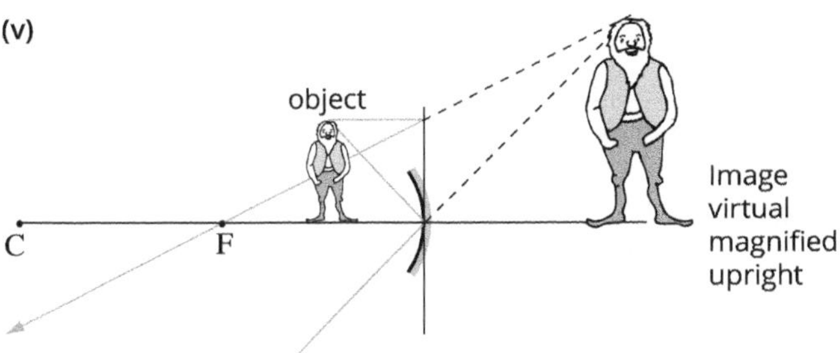

FIGURE 3.20 Images formed by a concave lens

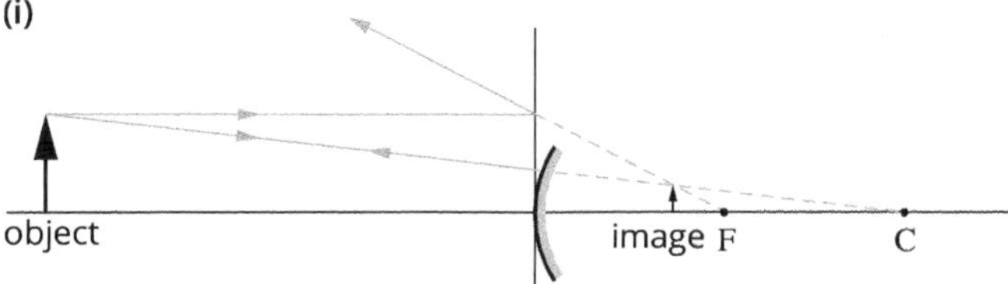

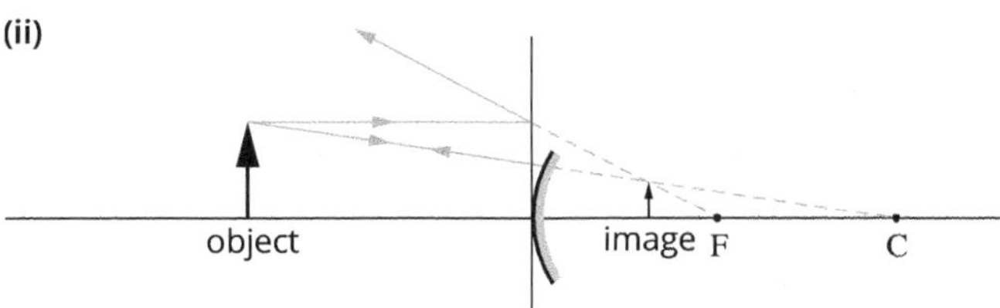

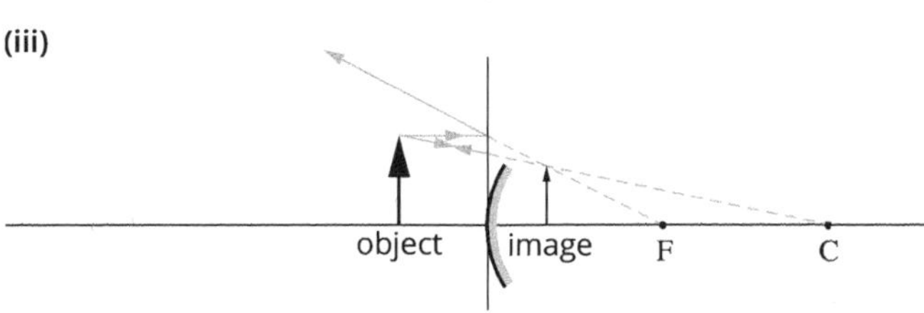

FIGURE 3.21 Images formed by a convex lens

$$M=\frac{h_i}{h_o}=\frac{v}{u}$$

where:

M is the magnification

h_i is the height of the image

h_o is the the height of the object

u is the the distance between the object and the mirror

v is the distance between the image and the mirror.

The units for height and distance must be the same.

Thermodynamics

Thermodynamics is the branch of physics that deals with heat and temperature (or more specifically, with thermal energy). Heat, temperature and thermal energy all refer to something slightly different, but they are all related. The behaviour of these quantities is governed by the four laws of thermodynamics.

HEAT AND TEMPERATURE

The kinetic particle model proposes that all matter is made of atoms or molecules (particles) that are in constant motion. Matter can exist in four phases (or states): solid, liquid, gas, and plasma. However, the three basic phases of matter (solid, liquid and gas) are the focus here. Solids have a fixed shape and volume as long as the temperature and pressure of the surroundings do not change.

- In solids, the forces of attraction and repulsion hold the particles in more or less fixed positions, usually in a regular arrangement or lattice. These particles are not completely still: they vibrate about average positions.
- In liquids there is still a balance of attractive and repulsive forces between particles, but the particles have more freedom to move around. Liquids have a fixed volume as long as the temperature and pressure of the surroundings do not change. Their shape is determined by the shape that contains them.
- In gases the particle speeds are high enough so that the attractive forces are not strong enough to keep the particles close together when they collide. The repulsive forces cause the particles to move off in other directions. Gases do not have a fixed volume; they fill the space that contains them.

Internal energy (or thermal energy) refers to the total kinetic and potential energy of the particles within a substance. **Heat** is the flow of thermal energy from one object to another. This process, known as heating, only transfers thermal energy from a hotter substance to a colder substance. For example, if an object has become hotter then it has gained heat energy. Conversely, if the object has become colder then it has lost energy. There are three different ways in which heat transfers conduction, convection and radiation.

Temperature describes 'how hot something is'. It is the property that determines the direction in which thermal energy is transferred to or from the object. Temperature is related to the average kinetic energy of the particles in a substance. The faster the particles move (or vibrate), the higher the kinetic energy of the substance and the greater the temperature.

 Temperature is measured in degrees Celsius (°C) or kelvin (K). Absolute zero is simply 'zero kelvin' (0 K) and is equal to −273.15°C. To convert from Celsius to kelvin, add 273.15; to convert from kelvin to Celsius: subtract 273.15.

When you heat an object, thermal energy is transferred from a hotter substance to a colder substance so that the entire system has no loss in energy. Heat is given the symbol Q. Because it describes energy it is measured in joules (J).

If no thermal energy flows between objects that are in contact with other, they are said to be in **thermal equilibrium**; that is, they have the same temperature.

The zeroth law of thermodynamics states that if objects A and B are both in thermal equilibrium with object C, then objects A and B are in thermal equilibrium with each other. A, B and C must be at the same temperature. A thermometer works by using the zeroth law; when you are in contact with a thermometer, your body heat is transferred to it until you and the thermometer have the same temperature.

The first law of thermodynamics states that energy simply changes from one form to another and the total energy in a system is constant. In other words, energy cannot be created or destroyed. Any change in the internal energy (ΔU) of a system is equal to the energy added by heating ($+Q$) or removed by cooling ($-Q$), minus the work done on ($-W$) or by ($+W$) the system: $\Delta U = Q - W$.

SPECIFIC HEAT CAPACITY

When heat is transferred to or from a system or object, the temperature change depends upon the amount of energy transferred, the mass of the material(s) and the specific heat capacity of the material(s):

 $Q = mc\Delta T$

where:

Q is the the heat energy transferred (in J)

m is the the mass of material being heated (in kg)

ΔT is the change in temperature (in °C or K)

c is the specific heat capacity of the material (in $\text{J kg}^{-1}\,\text{K}^{-1}$).

A substance will have a different specific heat capacity in different states (solid, liquid or gas).

A high specific heat capacity means that an object will absorb or release thermal energy at a slower rate. Objects with a low specific heat capacity will absorb or release thermal energy quickly.

LATENT HEAT

When a solid material undergoes a change of state (also called a phase change), energy is needed to overcome the attractive forces between the particles so that they can separate.

It is important to understand that the addition of thermal energy does not increase the kinetic energy during a change of state, so the temperature does not increase. This is shown at the 'ice melting' and 'water boiling' stages in Figure 3.22.

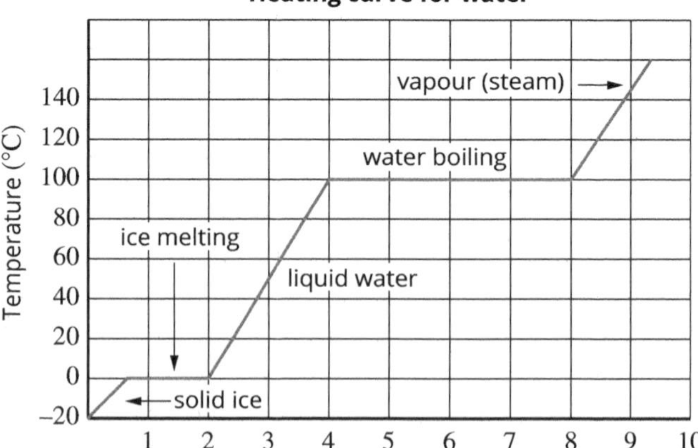

FIGURE 3.22 A heating curve for water, showing the constant temperatures during changes of state

The **latent heat** of a substance is the energy required to change the state of 1 kg of the substance at a constant temperature. There are two types of latent heat: latent heat of fusion and latent heat of vaporisation.

 ISBN 978 1 4886 1935 9

In general, for any mass of material the energy required (or released) is:

$Q = mL$

where:

Q is the energy transferred (in J)

m is the mass (in kg)

L is the latent heat (in $J\,kg^{-1}$).

The latent heat of fusion, L_{fusion}, is the energy required to change 1 kg of a material between the solid and liquid states. The latent heat of vaporisation, L_{vapour}, is the energy required to change 1 kg of a material between the liquid and gaseous states. The latent heat of fusion of a material will be different (and usually less than) the latent heat of vaporisation for that material.

Molecules in a liquid have varying speeds. The molecules travelling at a high enough speed can escape the liquid this is known as evaporation. Evaporation is when a liquid turns into gas at room temperature. The temperature of the liquid falls as this occurs. The rate of evaporation depends on the volatility, temperature and surface area of the liquid and the presence of a breeze.

CONDUCTION

One of the way heat is transferred is by conduction. **Conduction** is heat transfer within a material or between materials without the overall transfer of the substance itself. All materials conduct heat, but this process is most significant in solids. Materials that readily conduct heat, such as metals and glass, are called good thermal conductors. Materials that conduct heat poorly, such as wool and wood, are called thermal insulators. Whether a material is a good conductor depends on the method of conduction:

- Heat transfer by molecular collisions alone occurs in poor to very poor conductors.
- Heat transfer by molecular collisions and free electrons occurs in good to very good conductors.

The rate of conduction depends on the temperature difference between two materials, the thickness of the material, the surface area, and the nature of the material.

The rate of conduction can be calculated using the following formula:

$$\frac{Q}{t} = \frac{kA\Delta T}{d}$$

where:

k is the thermal conductivity (in $W\,m^{-1}\,K^{-1}$)

A is the surface area (in m^2)

ΔT is the change in temperature (in °C)

d is the thickness of the material (in m).

The thermal conductivity depends on the nature of the material. The unit for conduction is $J\,s^{-1}$, which is equal to 1 W.

It is possible to use this formula to calculate how much energy is transferred through the insulation in your home. For example, if you assume the insulation is 15 cm, the inside temperature is 20°C (5 K), and the outside temperature is 5°C, and taking the thermal conductivity of fibreglass to be $0.04\,W\,m^{-1}\,K^{-1}$, then the rate of energy lost through an area of 1 m is:

$$\frac{Q}{t} = \frac{kA\Delta T}{d} = \frac{0.04 \times 1 \times 15}{0.15}$$
$$= 4\,W$$

CONVECTION

Convection is the transfer of heat within a liquid or a gas as a result of the physical movement of matter. Unlike the other two types of heat transfer, it involves the mass movement of particles within a system. A convection current forms when there is warm fluid rising and cool fluid falling, caused by a difference in density. This movement mixes the high and low energies until thermal equilibrium is reached. An example of convection is thermal air currents in the atmosphere, which are used by glider pilots to gain altitude.

RADIATION

Any object with a temperature greater than absolute zero emits thermal energy by **radiation**. Radiant transfer of thermal energy from one place to another occurs by means of electromagnetic waves. An example of where radiation can be felt is sitting around a fire (Figure 3.23)—you don't have to touch the flame to feel the heat.

When electromagnetic radiation strikes an object, some is reflected, some is absorbed, and some may be transmitted.

The rate of emission or absorption of radiant heat will depend upon several factors:

- the temperature difference between the object and the surrounding environment
- the surface area and surface characteristics of the object
- the wavelength of the radiation.

Dark objects are good radiators and absorbers, whereas shiny and light-coloured objects are poor radiators and absorbers but good reflectors.

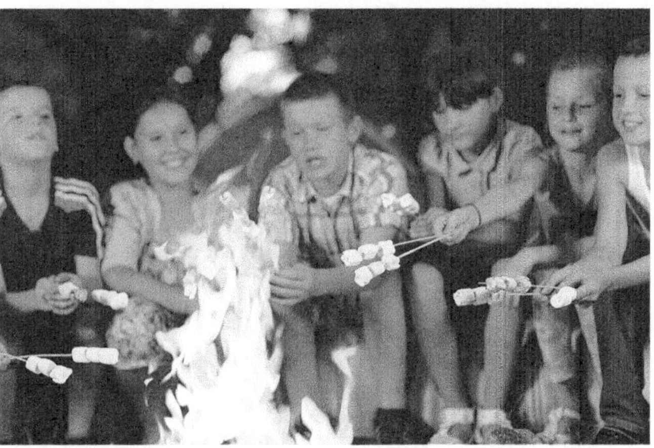

FIGURE 3.23 Heat from a fire is an example of radiation.

WORKSHEET 3.1

Knowledge review—waves and thermodynamics

1 Explain why the sound from a lightning strike is always heard after it is seen.

2 The speed of sound in air is about 330 m s^{-1}. If the thunder created by a lightning bolt is heard 12 seconds after the lightning is noticed, how far away was the lightning?

3 What is the difference between transverse and longitudinal waves?

4 Why do movie theatres often have thick curtains around the walls?

5 Put the following radiation types into order of increasing wavelength.

green infrared radio ultraviolet blue red x-rays microwaves

6 When a mirror is placed vertically in the middle of a letter, some capital letters such as A look the same.

a Which other letters would look the same in the mirror?

b Which letters would look the same if the mirror was placed horizontally half way up the letter?

7 What is the temperature, in degrees Celsius, of melting ice? What is it in kelvin?

8 For each of the situations in the table below, choose the correct name from the following list of words:

condensation, transformation, freezing, boiling, combustion, melting

Situation	Name
Solid changes to liquid	
Gas changes to liquid	
Liquid changes to solid	
Liquid changes to gas	

 ISBN 978 1 4886 1935 9

WORKSHEET 3.2

Waves around us

Waves are everywhere and have many shapes and forms. Vibrating objects transfer energy through waves travelling outwards. Other forms of waves transfer energy in the form of electromagnetic radiation.

Work through the following questions and activities to review your understanding of wave properties.

1 Write a single, short sentence that summarises the key properties of a wave.

____

____

2 Tick the appropriate box in the table below to categorise each of these examples as mechanical (M) or electromagnetic (E).

Wave	M	E
sound		
vibration		
radio		
visible light		
surf break		
microwave		
x-ray		
Mexican wave		
earthquake		
pulse in slinky		

Use the following displacement–time graphs of several transverse waves to answer Questions 3 to 5.

A

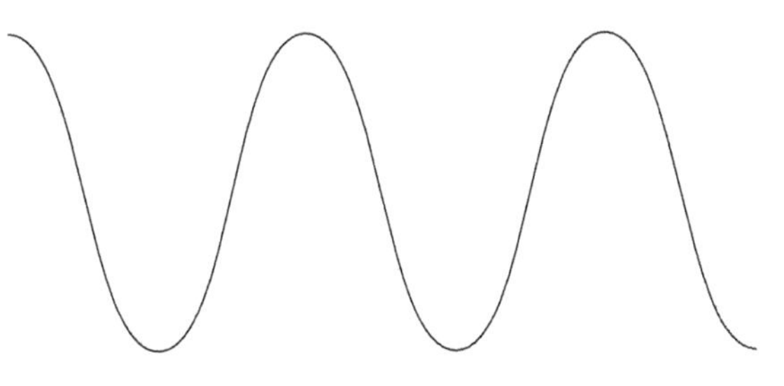

B

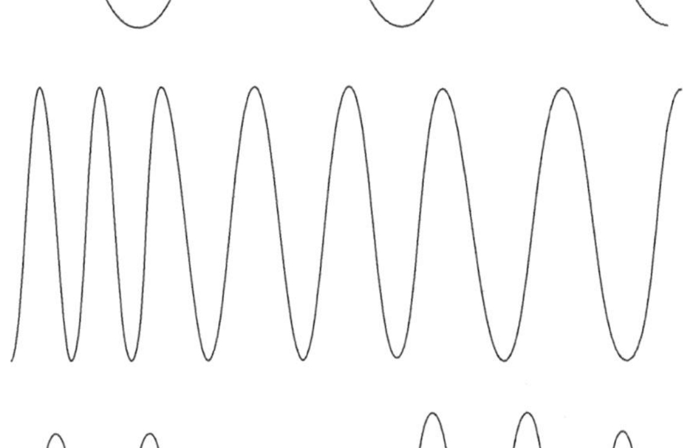

C

D

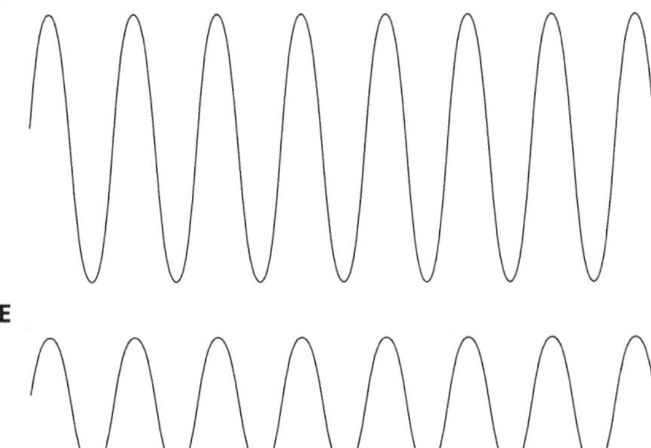

E

vertical scale: 1 cm = 1 cm
horizontal scale: 1 cm = 1 s

3 Which waves have the same maximum amplitude? What is the amplitude of these waves?

4 Which waves have the same frequency? What is that frequency?

5 Which wave has a changing frequency and constant amplitude?

6 The diagram below shows a transverse wave and a longitudinal wave.

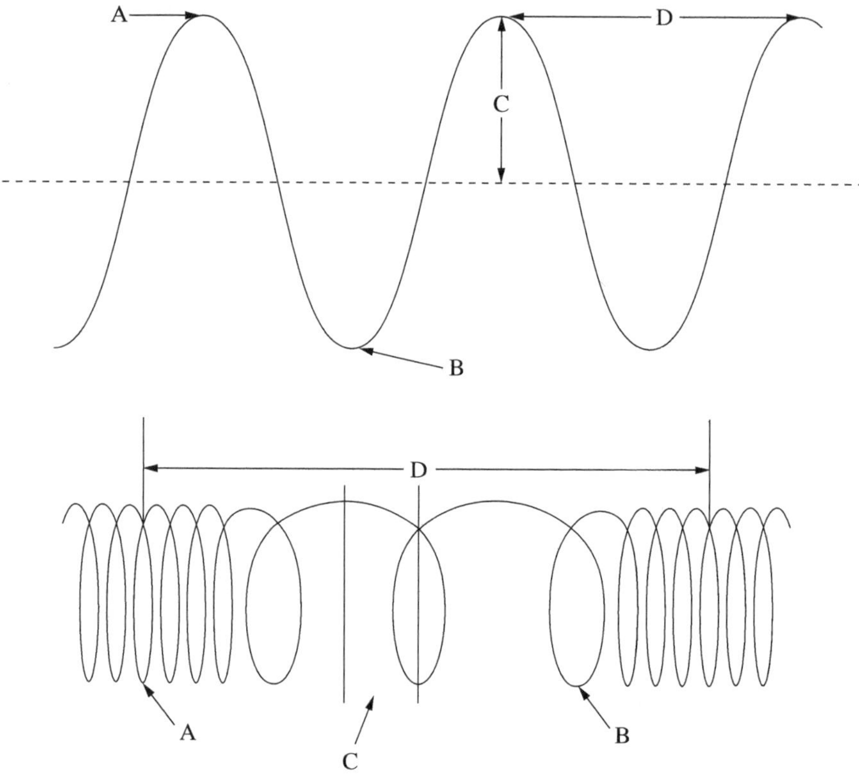

a Using the labelled key, identify the letter that corresponds to each key feature of each wave. Write your answers in the table below.

b Using a ruler and a scale of 1 cm = 1 cm, measure the amplitude and wavelength of each wave. Write your answers in the table below.

Key	Transverse wave	Size (cm)	Longitudinal wave	Size (cm)
A				
B				
C				
D				

RATING MY LEARNING	My understanding improved	Not confident ◄──► Very confident ○ ○ ○ ○ ○	I answered questions without help	Not confident ◄──► Very confident ○ ○ ○ ○ ○	I corrected my errors without help	Not confident ◄──► Very confident ○ ○ ○ ○ ○

ISBN 978 1 4886 1935 9

WORKSHEET 3.3

Interpreting graphs of waves

The following graphs show the displacement of two different waves.

Wave A shows the variation with distance (in m) of the displacement (in cm) of a transverse wave. The frequency of this wave is 5.0 Hz and the wave is moving to the right.

Wave B shows the variation with time (in s) of the displacement (in cm) of a longitudinal wave. The wavelength is 0.45 m.

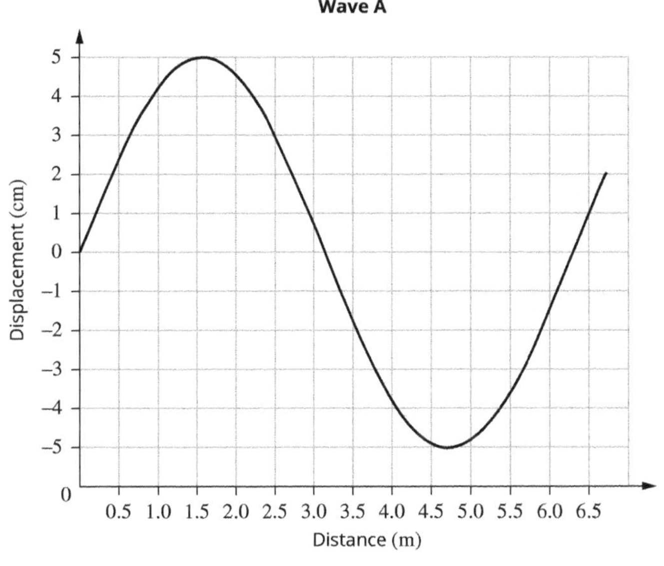

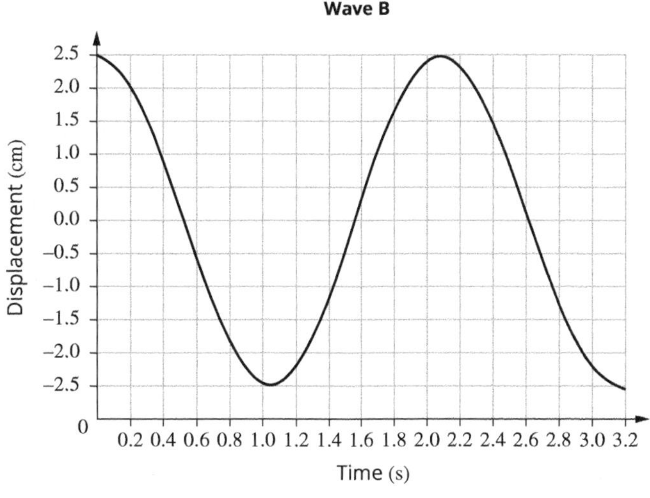

1 What is the amplitude of each wave?

2 Calculate the speed of Wave A.

3 Mark a point on the waves where the velocity of a particle in the wave is at a maximum.

4 Calculate the period for both waves.

5 What is the wavenumber for each wave?

RATING MY LEARNING	My understanding improved	Not confident ◄——► Very confident ○ ○ ○ ○ ○	I answered questions without help	Not confident ◄——► Very confident ○ ○ ○ ○ ○	I corrected my errors without help	Not confident ◄——► Very confident ○ ○ ○ ○ ○

WORKSHEET 3.4

The speed of sound by clap and echo

This is an individual or small group practical activity for home or study lesson. Allow around 10 minutes to collect your data.

MATERIALS

- sticks or tins to bang together
- 30 metre measuring tape
- stopwatch
- thermometer or weather app displaying the local temperature

PROCEDURE

1 Locate a large, open area with a large wall or rock face at one end. A school sports stadium is ideal. Measure out a distance of at least 30 metres from the wall, and preferably 100 metres or more. The longer the distance, the more reliable your final results will be.

2 Clap two pieces of wood together (for example, the flat sides of two rulers) or two tins (e.g. old paint tins) to produce a loud sound, and listen for the echo. Then clap again, allowing the same delay since the first echo as there was between the original clap and its echo. With a little practice, you should be able to establish a regular rhythm: clap – echo – clap – echo.

3 A second student should wait until the rhythm is well established and regular then start a stopwatch on their/your device at one of the claps. Count a group of about 50 claps then stop the stopwatch. Record the time and the number of claps. Repeat 3 or 4 times for a good average.

Distance from the wall: ________ m, Number of claps timed: ________,
Air temperature at time of activity: ________°C

Time recorded for each trial: 1:________ s; 2: ________ s; 3: ________ s;
4: ________ s

Average time: ________ s

1 What is the relationship between velocity, displacement and time? (Recall this from Module 1: Dynamics).

____

2 Based on your average time for *n* claps, what is the time for the sound to travel to the wall and back once?

____

3 Calculate the distance travelled by the sound in that time.

____

4 Calculate the speed of sound from your data.

____

5 The accepted value for the speed of sound is $v \approx 331 + 0.6T\,\text{m s}^{-1}$, where T is the air temperature. This is around $340\,\text{m s}^{-1}$ at 20°C. How does your answer compare?

____

6 There are many practical scenarios in which knowing the speed of sound can be useful in estimating distance; this is basically the reverse of this activity.

If a flash of lightning was followed by a thunder clap 4.5 seconds later, how far away was the lightning? Assume the temperature was approximately 28°C.

____

____

RATING MY LEARNING	My understanding improved	Not confident ◄──► Very confident ○ ○ ○ ○ ○	I answered questions without help	Not confident ◄──► Very confident ○ ○ ○ ○ ○	I corrected my errors without help	Not confident ◄──► Very confident ○ ○ ○ ○ ○

ISBN 978 1 4886 1935 9

WORKSHEET 3.5

Shifting it up—the Doppler effect

1 On the following diagram, show the change in the distribution of the wave fronts when:

a both observers and the ambulance (the sound source) are stationary

b the observers are stationary and the ambulance is moving to the left.

The ambulance is moving to the left at 72 km h^{-1}. The ambulance siren sounds two alternating tones of 960 Hz and 770 Hz. These are repeated every 1.3 seconds. Assume that the speed of sound at the prevailing temperature was 340 m s^{-1}.

2 What two apparent frequencies will be heard by the stationary observer that the ambulance is moving toward (standing on the left)?

3 What two apparent frequencies will be heard by the stationary observer that the ambulance is moving away from (standing on the right)?

4 Will the time between the tones that the observers hear also be changed? Explain. (If there is a change, calculate the time between tones that each observer will hear.)

5 Assuming the car is travelling at 72 km hr^{-1} and the ambulance is stationary, what would be the apparent frequencies of the two tones as the car is approaching the ambulance?

6 The Doppler effect is not restricted to affecting only the apparent frequencies of sound waves. The frequency of light from distant galaxies is affected by the Doppler effect. What does this observation tell us?

7 Some galaxies appear to have a 'red shift' in colour, so that the light they produce seems redder once it reaches Earth. Explain why this happens and what it tells us about the relative movement between these galaxies and the Earth.

RATING MY LEARNING	My understanding improved	Not confident ○ ○ ○ ○ ○ Very confident	I answered questions without help	Not confident 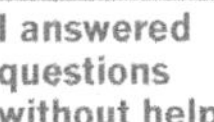 ○ ○ ○ ○ ○ Very confident	I corrected my errors without help	Not confident 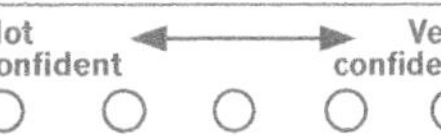 ○ ○ ○ ○ ○ Very confident

WORKSHEET 3.6

Lens diagrams

For each of the following scenarios, draw the ray diagram to find the resulting image, and describe the image (real or virtual, inverted or upright, magnified or reduced) by circling the correct options that apply for each diagram.

Diagram 1

Image: real/virtual inverted/upright magnified/reduced/same

f *f*

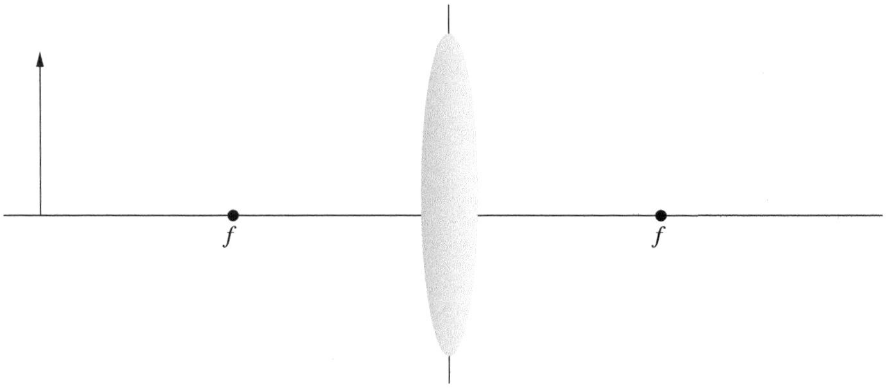

Diagram 2

Image: real/virtual inverted/upright magnified/reduced/same

f *f*

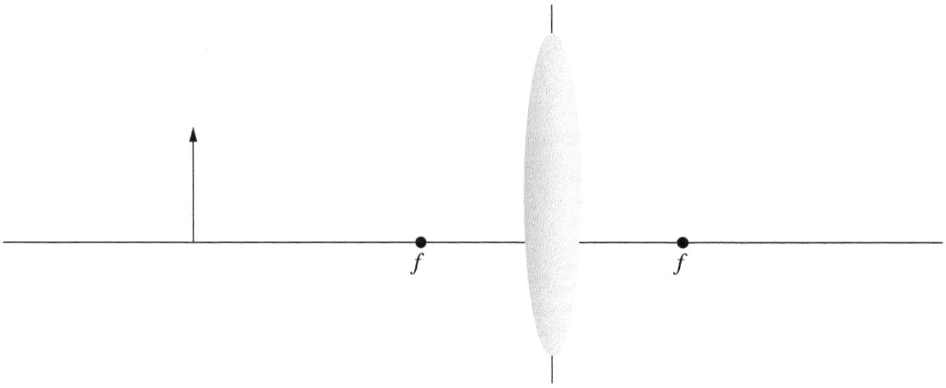

Diagram 3

Image: real/virtual inverted/upright magnified/reduced/same

f *f*

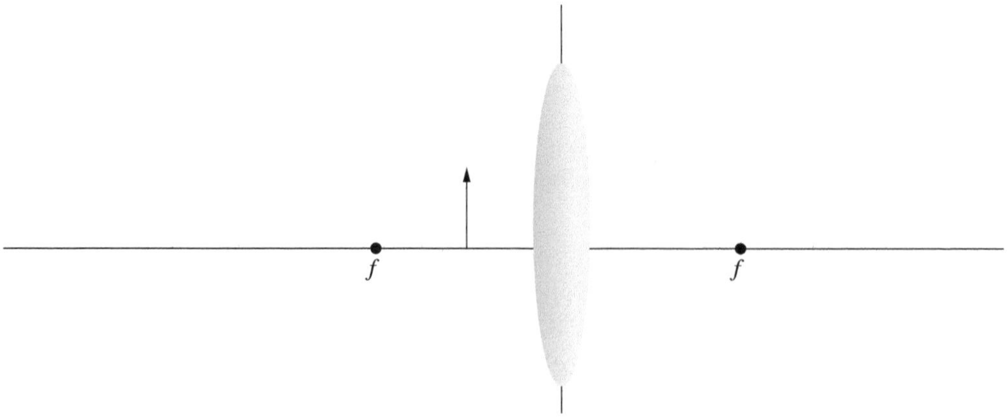

 ISBN 978 1 4886 1935 9

Diagram 4

Image: real/virtual inverted/upright magnified/reduced/same

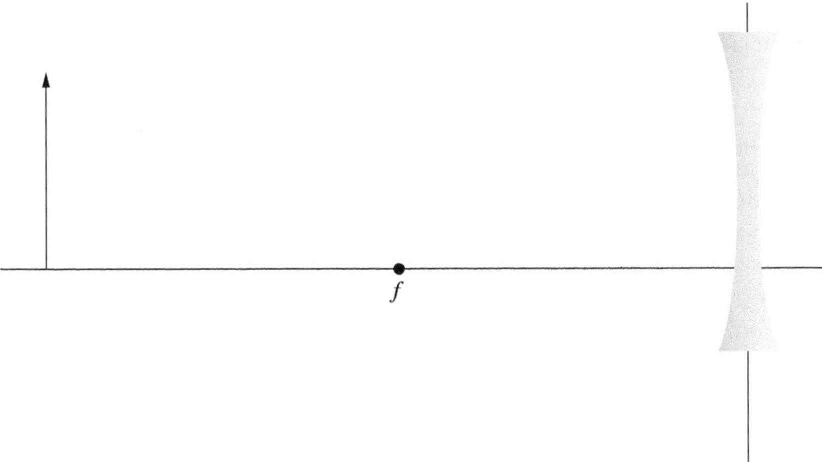

Diagram 5

Image: real/virtual inverted/upright magnified/reduced/same

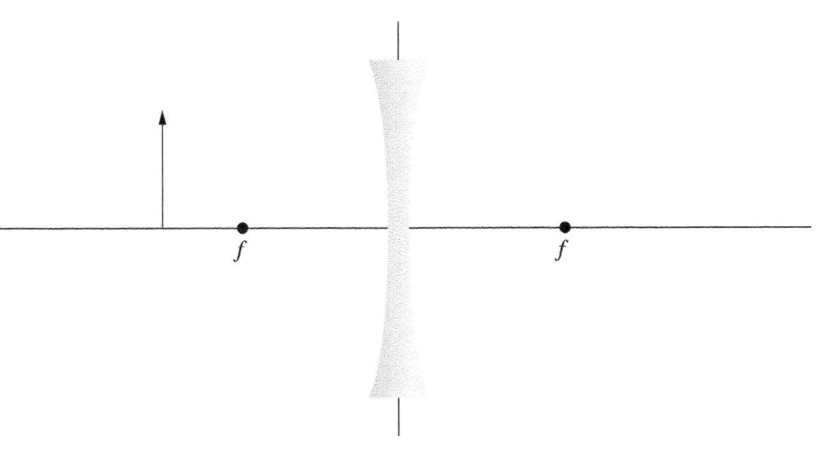

Diagram 6

Image: real/virtual inverted/upright magnified/reduced/same

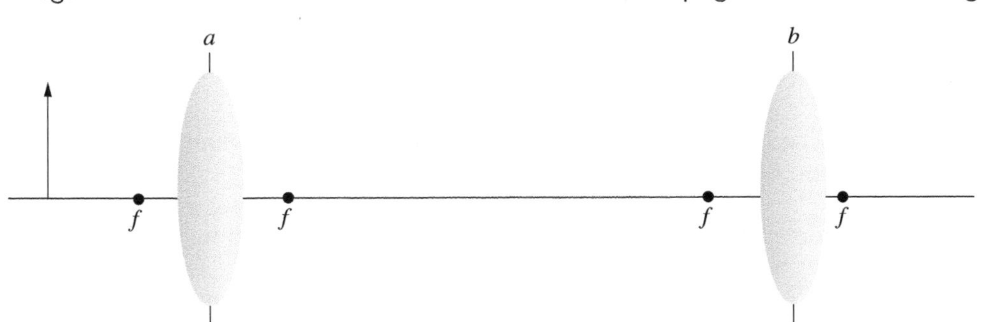

RATING MY LEARNING	My understanding improved	Not confident ⟷ Very confident ○ ○ ○ ○ ○	I answered questions without help	Not confident ⟷ Very confident ○ ○ ○ ○ ○	I corrected my errors without help	Not confident ⟷ Very confident ○ ○ ○ ○ ○

WORKSHEET 3.7

Designing a musical instrument

'Without music, life would be a mistake.' — Friedrich Nietzsche

'If I were not a physicist, I would probably be a musician. I often think in music. I live my daydreams in music. I see my life in terms of music.' — Albert Einstein

Almost all musicians create music by making standing waves in their instrument. The guitar player makes standing waves in the strings of the guitar, and the drummer does the same in the skin of the drumhead. The trumpeter and flautist both create standing waves in a column of air. When the desired note, or tone, is played on a musical instrument, there are always additional tones produced at the same time. The frequency of the desired note is known as the fundamental frequency or first harmonic, which is caused by the first mode of vibration, but many higher modes of vibration always naturally occur simultaneously. Higher modes are simply alternative, higher-frequency vibrations of the musical instrument medium. The frequencies of these higher modes are known as overtones.

There is a very simple relationship between the overtones and the fundamental frequency. Harmonics are overtones that are a simple integer multiple of the fundamental frequency. For example, if a plucked string produces a frequency of 110 Hz, multiples of 110 Hz will also occur at the same time: 220 Hz, 330 Hz, 440 Hz, etc. will all be present, although not all with the same intensity. The fundamental frequencies of each note, and all of their overtones, combine to produce an instrument's sound spectrum.

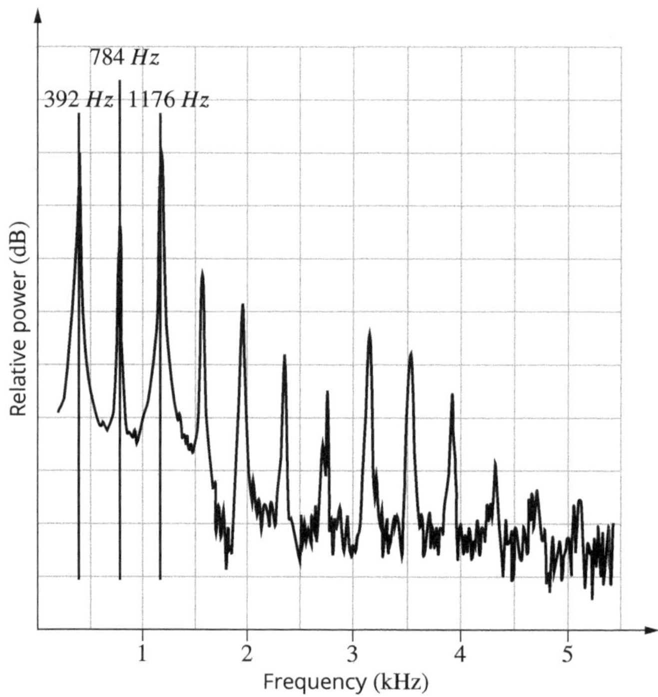

An octave on a musical scale represents a doubling of the frequency. In the sound spectrum of the flute pictured above, the first overtone at 784 Hz would therefore be the next highest G, or G_5.

By adopting this equal, whole number spacing, notes one octave apart are harmonious when played together. Scales are then broken down into intervals depending on how many tones or notes are required in between. Two constraints are that the frequencies chosen should be evenly spaced and that the frequencies will sound good when played together. Musical instruments are then designed in a way that enables all the notes within one or more octaves to be played individually or together.

Modern instruments are designed to play notes on the 'equal tempered chromatic scale'. This scale has each octave broken down into a number of equally spaced intervals called semitones.

Answer the following questions, basing your answer on this scale and referring to a reliable source.

 ISBN 978 1 4886 1935 9

WORKSHEET 3.7

1 Middle C, also known as C_4, is the centre C of a piano keyboard. It has a frequency of 256 Hz. What is the frequency of C_2 and C_5, the second and fifth C on the keyboard, respectively?

2 How many intervals is the equal tempered chromatic scale broken into?

3 The ratio of the frequencies of any two adjacent notes on this scale is the same. What is the ratio?

4 Using the ratio you have found, work out the frequency of each note, or semitone, between C_4 and C_5 based on a middle C_4 of 256 Hz. Identify each frequency using the standard alphabetic symbol used in modern music. Each note in the table below is a semitone higher than the preceding note. The symbol # means 'sharp'.

C_4	C#	D	D#	E	F	F#	G	G#	A	A#	B	C_5

5 Draw the first mode of vibration of a wave produced by a stringed instrument. What is the relationship for determining the frequency played from the length of the string?

6 Draw the string vibrating in the 2nd, 3rd and 4th modes of vibration. Derive the equation for the frequency for each mode of vibration.

7 Examine the pattern between each of the modes of vibration. Based on this pattern, write the equation for the *n*th mode of vibration.

8 The frequency of the first C on a C-tuned guitar, C_2, is 65.406 Hz. Use this frequency and the equal tempered scale to calculate the frequencies of all the unfretted strings. (Unfretted means not being shortened by pressing on the string with a finger.) The resulting notes can be described as C-F-A#-D#-G-C or (more often) C-F-B♭-E♭-G-C.

9 The speed *v* of a wave in a guitar string is influenced by three variables. Identify the variables and use them to write an expression for *v*.

10 By holding a finger against a string you decrease its length and change the frequency. This is called fretting. At what length along the E_2 string (82.4 Hz) would it need to be fretted in order for it to give the same frequency as the A_2 string? The total string length is 0.65 m.

11 Calculate the length the E_2 and A_2 strings would need to be fretted in order to have the frequency of D_3.

12 A_2 is __________ semitones higher than E_2.

13 Use your answers to Questions 11 and 12 to make a statement about how you think the placement of frets is determined. As evidence for your hypothesis, include information from your answers to Questions 11 and 12.

RATING MY LEARNING	My understanding improved	Not confident ◄——► Very confident ○ ○ ○ ○ ○	I answered questions without help	Not confident ◄——► Very confident ○ ○ ○ ○ ○	I corrected my errors without help	Not confident ◄——► Very confident ○ ○ ○ ○ ○

 ISBN 978 1 4886 1935 9

WORKSHEET 3.8

Communicating is 'critical'

Modern communication systems such as the National Broadband Network use optical fibres rather than copper or wire to transmit thousands of signals down one fibre at the speed of light. The basis for the success of fibre-optic cable is the total internal reflection of light within the cable.

Light travels down a fibre-optic cable by reflecting repeatedly off the walls at a shallow angle, less than the critical angle for the material. This is the core of the fibre. Bonded around the core is another layer called the cladding. The cladding ensures that the light stays within the core. It does this because it is made of a different type of glass with a lower refractive index than the core.

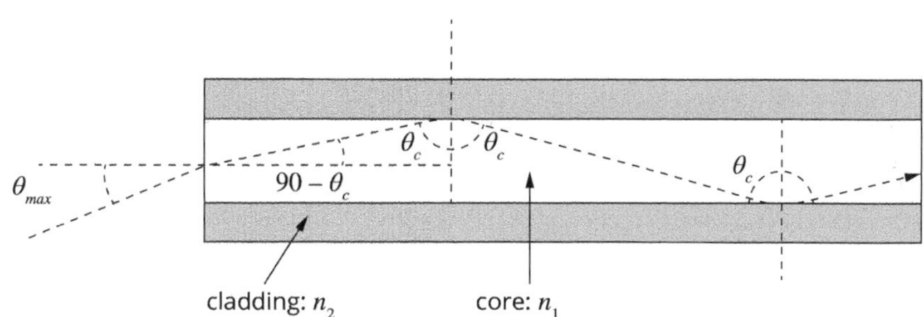

1 A particular glass fibre-optic has an index of refraction of 1.48. The cladding material has a lower refractive index of 1.46. Find the sharpest angle at which the fibre optic can be bent so the light remains in the cable.

2 Another fibre-optic cable has a core with a refractive index of 1.65. Around this core the cladding has a refractive index of 1.35. Calculate the critical angle for this cable.

3 For the cables in questions 1 and 2, calculate θ_{max} (referred to as the 'cut-off angle') so that the light will be completely internally reflected. Assume that the light is entering from air, $n = 1$.

4 Which cable would be the most flexible for use in a general-purpose communications network? Justify your answer with reference to your answers to the previous questions. What other considerations would need to be given to the choice of cable?

5 The following three diagrams show light travelling inside a fibre-optic cable.

The critical angle for the cable is 42°. Draw the path that the light would take in each case.

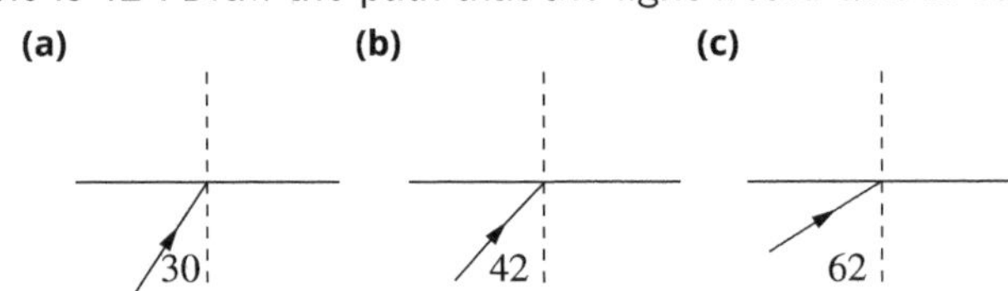

For the following questions, three materials (A, B and C) are being considered for the core and cladding of an optical fibre. For material A, $n = 1.20$; for material B, $n = 1.35$; and for material C, $n = 1.45$.

6 Find the critical angle if material B is used as the core and material A is the cladding.

7 Find the critical angle if material C is used as the core and material B is the cladding.

8 What combinations of materials for the core and cladding are not possible?

RATING MY LEARNING	My understanding improved	Not confident ◄──► Very confident ○ ○ ○ ○ ○	I answered questions without help	Not confident ◄──► Very confident ○ ○ ○ ○ ○	I corrected my errors without help	Not confident ◄──► Very confident ○ ○ ○ ○ ○

 ISBN 978 1 4886 1935 9

WORKSHEET 3.9

Climate in the balance—the role of oceans and air in climate control

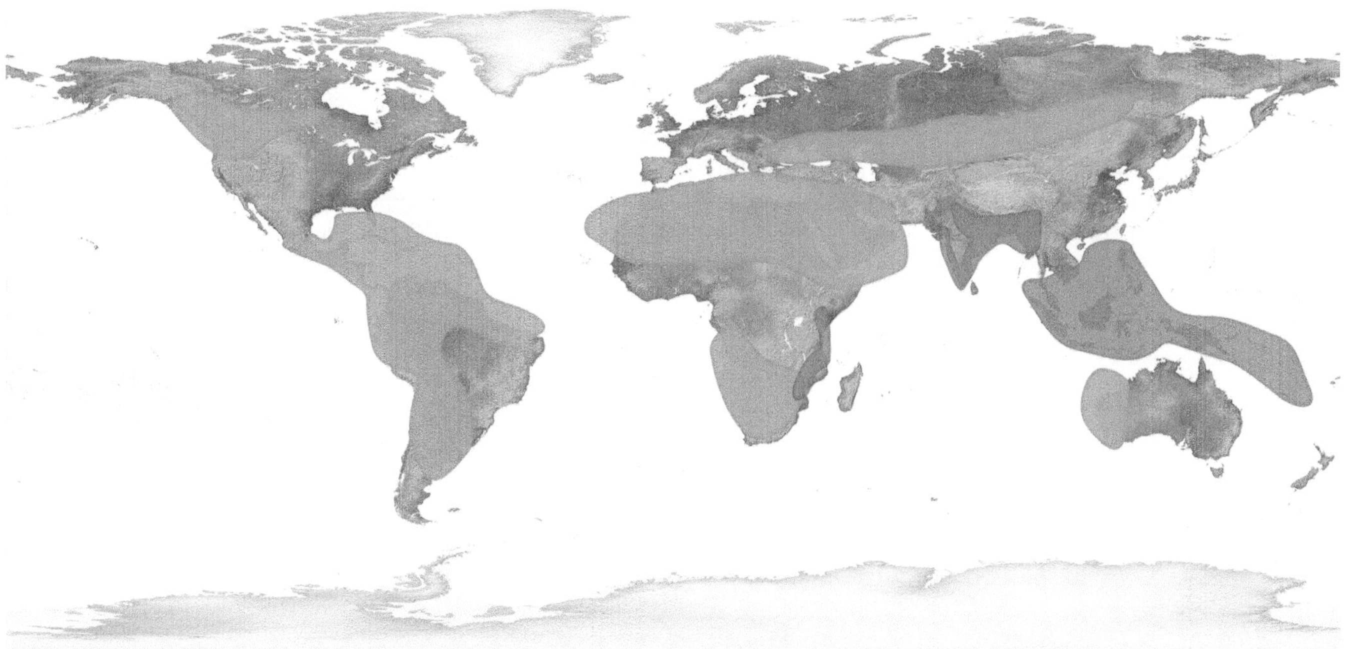

Water has a surprisingly high specific heat capacity, 10 times higher than most metals. The specific heat capacity of water is in fact higher than for all but a very few uncommon materials. As a result, water makes a very useful cooling and heat storage agent; for example, it is used in generator cooling towers and car engine radiators.

Life on Earth depends on the specific heat capacity of water. About 70% of the Earth's surface is covered by water, and our oceans are the primary regulator of global climate through absorption, storage and transport of heat energy.

To a large degree, the ability for us to understand what might happen to the Earth's climate in the future depends on understanding the role of the oceans in maintaining and moderating climate change.

The following questions look at how much energy can be stored and transferred by the oceans.

1 How much energy does it take to heat 1 litre of water by 1°C?

2 The oceans around the equator are exposed to direct sunlight for around 12 hours every day. If the temperature of the top 1 cm of water over 1000 km^2 increases by an average 2°C from 26°C to 28°C in one day, how much heat energy has the water absorbed?

3 If this heat energy were to be released into the atmosphere overnight, what volume of air would be warmed by 1°C? Assume that all the energy was transferred. You can use the following information:

1 kg of dry air at standard temperature and pressure has a volume of 0.8562 m^3

c for air at approximately 26°C is 1.005 $J\,kg^{-1}\,K^{-1}$

4 The total mass of the Earth's atmosphere has been estimated at 5.1×10^{18} kg. If the atmosphere had a uniform composition, what would be the change in temperature for the whole atmosphere from the release of the energy mentioned in Question 2?

5 Since 1950, the temperature of the top 300 mm of the ocean's surface has risen by about 0.3°C. Over the last 50 years, more than 90% of the excess heat energy from human activities has been stored as heat energy in the Earth's oceans, amounting to a total of 2×10^{23} J. If this energy were to be suddenly released, what would be the average temperature change in the Earth's atmosphere?

6 Assume now that this heat energy was instead used to heat ice from ice at 0°C to water at 0°C, what mass of ice would be melted? Assume that L_{fusion} for water is 3.34×10^5 J.

7 There are a number of moderating influences that mean these massive changes are not likely to happen. From your understanding of the movement of heat energy, explain the mechanisms behind heat transfer in the Earth's atmosphere.

RATING MY LEARNING	My understanding improved	Not confident ◄──► Very confident ○ ○ ○ ○ ○	I answered questions without help	Not confident ◄──► Very confident ○ ○ ○ ○ ○	I corrected my errors without help	Not confident ◄──► Very confident ○ ○ ○ ○ ○

 ISBN 978 1 4886 1935 9

WORKSHEET 3.10

Literacy review—describing energy and waves

1 Use the following word list to complete the text below.

sound	longitudinal	energy	parallel
matter	pulse	mechanical	perpendicular
transverse	guitar string	periodic	

A wave may be a single ________________ or it may be continuous or ________________. A wave only transfers ________________ from one point to another. There is no net transfer of ________________. ________________ waves can be either ________________ or ________________. In a transverse wave, the oscillations are ________________ to the direction in which the wave energy is travelling. A vibrating ________________ is an example of a transverse wave. In a ________________ wave the oscillations are ________________ to the direction the wave energy is travelling. ________________ is an example of a longitudinal wave.

2 Use the following word list to complete the text below.

transmitted	fixed	reflected	phase
incident	absorbed	boundary	free

A wave reaching the ________________ between two materials in which it can travel is partly ________________ and partly ________________, and may be partly ________________. This incoming wave can be split into three different waves: an ________________ wave, a ________________ wave and a ________________ wave.

The reflected wave responds in one of two ways when it reaches a boundary:

Waves reflect with a 180° phase change from __________ boundaries.

Waves reflect with no phase change from __________ boundaries.

3 Use the following word list to complete the text below.

speed	medium	frequency	refraction
wavelength	propagation	speed	

When waves pass from one ________________ to another, further properties of waves can be observed. ________________ is the same, but ________________ and ________________ change. ________________ is the change in direction of ________________ of a wave that occurs when there is a change in ________________ of the waves.

4 Using the Key Knowledge section of this module, your textbook or another suitable source, define the following key terms from your study of sound waves.

Diffraction ________________________________

Resonance ________________________________

Superposition ________________________________

Harmonic ________________________________

Interference ________________________________

Intensity ________________________________

Pitch ________________________________

Loudness ________________________________

Standing wave ________________________________

ISBN 978 1 4886 1935 9

5 Complete the following paragraph relating to specific and latent heat from the word list provided.

phase	specific heat capacity	energy	transferred
temperature	latent	energy	

When heat is ________________ to or from a system or object, the ________________ change depends upon the amount of ________________ transferred, the mass of the material and the ________________ of the material. When a solid material changes state, ________________ is needed to separate the particles by overcoming the attractive forces between the particles. This is known as a ________________ change. The energy required to do so is referred to as ________________ or hidden heat.

6 From the Key Knowledge section of this module, your textbook or another suitable source, define each of the following means of heat transfer.

Conduction ____

Convection ____

Radiation ____

RATING MY LEARNING	My understanding improved	Not confident ◄──► Very confident ○ ○ ○ ○ ○	I answered questions without help	Not confident ◄──► Very confident ○ ○ ○ ○ ○	I corrected my errors without help	Not confident ◄──► Very confident ○ ○ ○ ○ ○

 ISBN 978 1 4886 1935 9

WORKSHEET 3.11

Thinking about my learning

After completing Module 3: Waves and thermodynamics, you should be able to describe, explain and apply the relevant scientific ideas. You should be able to work with data, to interpret, analyse and evaluate it.

1 The table below lists the key knowledge covered in this module. Read each and reflect on how well you understand each concept. Rate your learning by shading the circle that corresponds to your level of understanding for each concept. It may be helpful to use colour as a visual representation. For example:

- green—very confident
- orange—in the middle
- red—starting to develop.

Concept focus	Rate my learning				
	Starting to develop ◄				► Very confident
Types of waves: longitudinal, transverse, mechanical and electromagnetic	○	○	○	○	○
Wave characteristics: velocity, frequency, period, wavelength, wave number and amplitude	○	○	○	○	○
Wave behaviour: reflection, refraction, diffraction and superposition	○	○	○	○	○
Resonance	○	○	○	○	○
Standing waves in strings and pipes	○	○	○	○	○
Sound phenomena including beats and the Doppler effect	○	○	○	○	○
Images formed in mirrors and lenses	○	○	○	○	○
Light behaviour, including intensity, Snell's law and dispersion	○	○	○	○	○
Kinetic particle theory	○	○	○	○	○
Specific heat capacity	○	○	○	○	○
Types of energy transfer: conduction, convection and radiation	○	○	○	○	○
Latent heat	○	○	○	○	○

2 List any specific concepts that you found challenging in this module.

3 Write down two different strategies that you will apply to help improve your understanding of these concepts.

PRACTICAL ACTIVITY 3.1

Waves in slinkies and ropes

Suggested duration: 30 minutes

INTRODUCTION

Particles affected by the motion of transverse waves move perpendicularly to the direction of the wave. Particles affected by a longitudinal wave move in the same direction as the wave's motion. The transfer of energy through either a longitudinal or a transverse wave is affected by the medium in which the wave travels.

MATERIALS

- 2 slinky springs with different diameters
- 2 metres of string or light rope
- light ropes or strings between 5 and 10 metres long, with various masses per metre

PURPOSE

To investigate some simple properties of longitudinal and transverse waves and the effect of the medium on the propagation of the waves.

PROCEDURE

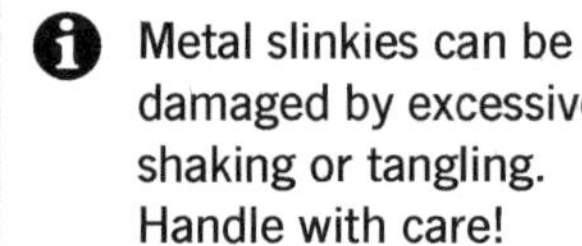

Metal slinkies can be damaged by excessive shaking or tangling. Handle with care!

Part A

1 Lie the larger, or heavier composition, slinky spring flat on the floor in a cleared space. With each end held by a person, stretch the slinky until it is taut but not over-stretched.

2 Generate a pulse by whipping one end of the slinky out and back along the floor and at right angles to the line of the slinky. Record what happens to the pulse as it travels down the slinky and reflects back from the fixed end.

3 Repeat with the smaller spring and record what happens. Pay particular attention to any difference between the movement and reflection of the pulse in the smaller and larger springs.

4 Now attach the short piece of string or light rope to the end of the larger diameter slinky. Once again stretch the slinky and the string out between two people without over-stretching the slinky. Generate a single pulse in the slinky and record what happens at the transition.

5 Remove the string and stretch out one of the slinkies between two people.

6 Create a longitudinal pulse by having one person move their end of the slinky sharply in and out in the direction of the slinky (longitudinally). Record what happens as the pulse travels down the slinky and reflects off the fixed end.

7 Now join the two slinkies together and stretch them out along the floor.

8 Keeping the end of the larger slinky fixed, generate a transverse pulse in the smaller diameter slinky.

 ISBN 978 1 4886 1935 9

9 Record and sketch what happens as the pulse travels along and through the transition between them.

10 Repeat the previous step, but this time generate the pulse in the larger slinky while keeping the end of the smaller slinky fixed.

11 Record and sketch what happens as the pulse travels along and through the transition between them.

Part B

1 Fix one end of a long rope above the floor. For example, tie it to the handle of a closed door.

2 Tie a small ribbon or rubber band anywhere along the rope (It does not have to be in the centre, but don't make it too close to either end.)

3 Have one student pick up the free end of the rope and take up the slack, without pulling it tight. Position another student level with the ribbon to observe its motion.

4 Flick the rope sharply up and down to produce a single transverse pulse that travels along the rope.

5 Record the movement of the ribbon.

6 Repeat steps 1 to 5 with different tensions on the rope.

7 Repeat steps 1 to 5 with ropes of varying mass and composition.

8 Now tie a loop at the end of the rope around a thin, upright object such as the leg of a desk. Once again, flick the rope to produce a single transverse pulse that travels along the rope, and record what happens when the wave is reflected from this free end.

CONCLUSION

1 In this practical activity, the waves generated have also been referred to as pulses. Why?

2 Compare the behaviour of a transverse wave and a longitudinal wave in the slinky as they were reflected back from the fixed end.

PRACTICAL ACTIVITY 3.1

3 Describe the behaviour of the transverse wave as it passed through the transition from the smaller slinky to the larger slinky.

4 Describe the behaviour of the pulse in the rope when it reached a fixed end.

5 Describe the behaviour of the pulse in the rope when it reached a free end.

6 Does the tension in a rope have any affect on the speed of a pulse travelling along it?

7 Is the speed of a pulse in a rope affected by the mass or diameter of the rope?

8 Using your observations of the ribbon on the rope, describe how the direction of travel of a particle in a transverse wave is different from the direction of propagation of the pulse itself.

9 List some examples of naturally occurring transverse and longitudinal waves.

10 The waves produced in this activity all depend on having a medium to travel in (a slinky or rope). What sort of wave does not need a medium in order to travel?

RATING MY LEARNING	My understanding improved	Not confident ◄—► Very confident ○ ○ ○ ○ ○	I answered questions without help	Not confident ◄—► Very confident ○ ○ ○ ○ ○	I corrected my errors without help	Not confident ◄—► Very confident ○ ○ ○ ○ ○

ISBN 978 1 4886 1935 9

PRACTICAL ACTIVITY 3.2

The speed of sound by resonance tube

Suggested duration: 40+ minutes

INTRODUCTION

Standing waves generated in a closed tube have a node at the closed end and an antinode at the open end. The simplest harmonic, or fundamental frequency, will have a wavelength four times the length of the effective air column.
In general:

$$\lambda_n = \frac{4l}{n} \text{ and } f_n = \frac{nv}{4l}$$

where:

λ_n is the wavelength (in m)

l is the effective length of the air column (in m)

f_n is the frequency of the harmonic (in Hz)

v is the velocity of the waves (in m s^{-1})

n is the number of the harmonic (1, 3, 5, ...)

(Only odd-numbered harmonics are possible in a closed air column.)

Rearranging the relationship between the frequency of a harmonic and the velocity of the sound in the particular medium provides a means of determining the velocity of sound. For $n = 1$:

$$f_1 = \frac{1 \times v}{4l_1} \rightarrow v = 4l_1f_1$$

For $n = 3$:

$$f_3 = \frac{3 \times v}{4l_3} \rightarrow v = \frac{4l_3f_3}{3}$$

PURPOSE

To determine the speed of sound in air using a resonance tube.

 Tuning forks are fragile and easily damaged. A light tap is all that is needed to start them vibrating.

MATERIALS

- tuning forks (or digital signal generator and speaker)
- clear glass or plastic tube at least 2 cm diameter and 1 metre long, sealed at one end so that it is watertight (tubes made from other materials can be used but must be waterproof)
- thermometer or temperature sensor (one per class is sufficient)
- water
- metre ruler

PROCEDURE

1 Fill the tube with water to about 2 cm from the open end.

2 Vibrate a tuning fork by flicking it with a finger or tapping it lightly against a solid surface, and hold it sideways over the mouth of the tube so that the vibrations from the tuning fork will travel down the length of the tube. Listen carefully for a louder sound to emerge from the tube. This is resonance. The length of the air column down to the water surface matches a resonant position. The effect is much like blowing over the end of a tube.

PRACTICAL ACTIVITY 3.2

 You will need a quiet room and patience to hear resonance. Resonance positions occur only at very specific lengths.

3. If a louder sound indicating resonance is not heard, pour a small amount of water out of the tube to lengthen the air column and try again.
4. Keep pouring out small amounts and trying again until resonance is clearly heard. Adjust the amount of water in the tube by adding or subtracting water to give the loudest sound close to this length.
5. Measure the length l_1, which is the distance from the open end of the tube down to the water surface, and record it in the table.
6. Pour water out of the tube to create an air column a little less than three times l_1.
7. Sound the same tuning fork over the mouth of the tube and listen again for resonance. Adjust the amount of water in the tube until resonance is heard clearly.
8. Measure length l_3 from the open end of the tube down to the water surface and record it in the table below.
9. Repeat steps 2 – 4 with tuning forks of different frequencies.
10. Measure the temperature in the room and record it in the table.

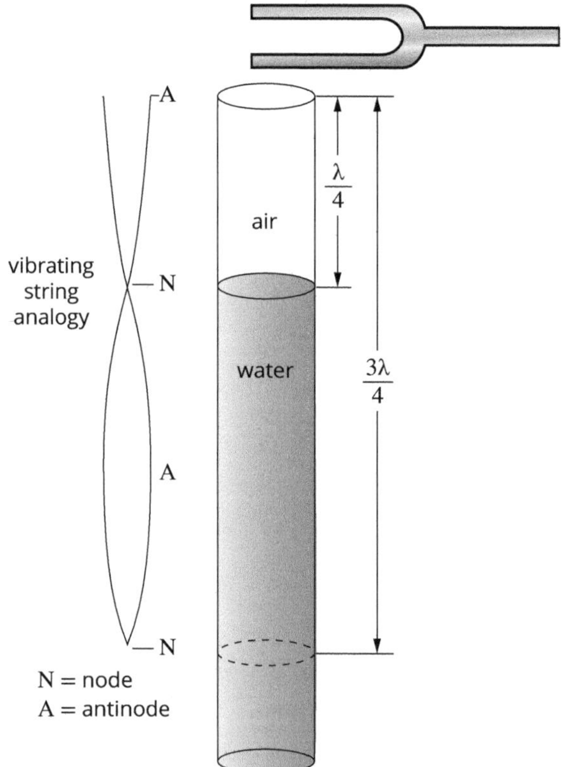

 ISBN 978 1 4886 1935 9

DATA AND ANALYSIS

Room temperature: _______°C

Frequency (f) (Hz)	l_1 (m)	l_3 (m)	$l_3 - l_1$ (m)	λ (m)	$\frac{1}{f}$ (s)	v (m s^{-1})
Average velocity (m s^{-1})						

Determining the speed of sound with single results

1 For each frequency for which two resonant positions at lengths l_1 and l_3 were recorded, calculate the wavelength using the formula $\lambda = 2 \times (l_3 - l_1)$. Record your calculations in the table.

Note: The air pressure of the standing wave will support the wave for a short distance outside the tube. Using the value of $l_3 - l_1$ eliminates this effect.

2 For each wavelength calculated in step 1, calculate the experimental speed of sound using $v = f\lambda$. Record your calculations in the table and calculate the average velocity from the results available.

Determining the speed of sound with class results

3 Graph l_1 against $\frac{1}{f}$. Because $l_1 \propto \frac{v}{f}$, the gradient of the graph can be used to determine the velocity.

4 Record your calculation of the velocity from the gradient of the graph and the value obtained.

CONCLUSION

1 Briefly describe the difference between a standing wave (like the one created in the tube) and a progressive wave.

2 Why is it important to record the temperature for this investigation?

3 How did the sound change when you used a tuning fork with different frequency?

4 What value did you obtain for the speed of sound using:

a your results?

b graphically using the class results?

5 Research the speed of sound at the temperature recorded in the room. Calculate the percentage difference between the accepted value and your experimental value. You may need the formula:

$$\text{Percentage difference} = \frac{\text{theoretical value} - \text{experimental value}}{\text{theoretical value}} \times 100$$

6 How do the experimentally determined results compare with accepted values? Comment with reference to the percentage difference from the accepted value.

7 Why was a graph used to find velocity from a class set of results?

RATING MY LEARNING	My understanding improved	Not confident ◄——► Very confident ○ ○ ○ ○ ○	I answered questions without help	Not confident ◄——► Very confident ○ ○ ○ ○ ○	I corrected my errors without help	Not confident ◄——► Very confident ○ ○ ○ ○ ○

 ISBN 978 1 4886 1935 9

PRACTICAL ACTIVITY 3.3

Image formation in lenses and mirrors

Suggested duration: 45 minutes

INTRODUCTION

Rays of light reaching a small mirror from a distant object can be considered to be parallel for all practical purposes. Parallel rays of light are focused at the focal point of a concave mirror. The image of a distant object will therefore be in focus at a point which is 1 focal length in front of the mirror.

The image of an object placed less than the focal length from the mirror will be located at a distance equivalent to the centre of curvature of the mirror.

The phenomenon of parallax can be used to determine the focal length of a lens.

When you look at the needle of the speedometer of a car from the driver's seat, the view from directly in front shows the correct, indicated speed. Viewing the needle at the same speed from the passenger's seat can suggest that the car is travelling faster. This is because the needle sits out from the dial, so when it is viewed from an angle it aligns with a higher number on the speedometer. This error is referred to as parallax error.

If the needle was in the same plane as the dial then no parallax error would occur and the indicated speed would be the same regardless of the position of the observer.

Parallax can be used to determine the focal length of a convex lens.

MATERIALS

- concave mirror of focal length 20 cm or less, with holder
- concave mirror of unknown focal length
- convex lens of unknown focal length
- plane mirror
- small screen of white plastic or cardboard
- pearlescent light globe and holder
- power supply for globe
- white image screen with 1 cm hole and cross-hairs (see diagram below)
- retort stand and clamp
- long pin or needle mounted in a small cork
- ruler

PURPOSE

To determine the focal length of a concave mirror and a convex lens.

A pearlescent globe is used for this activity to avoid any confusion between the image of the cross-hairs in the screen and the filament of the globe.

Alternatively, use an LED lamp or torch.

PROCEDURE

Part A: Concave mirror

Method 1: Focusing a distant object

1 Face a concave mirror toward a window where there are distant objects visible outside. Hold the white plastic or cardboard screen in a shaded area in front of the mirror.

2 Adjust the position of the screen or mirror until sharp images of distant objects outside the window are formed on the screen.

3 Record the nature of the image (inverted or upright, magnified or reduced) in a table.

4 Measure the distance between the screen and the mirror. This will be a good estimate of the focal length. Record the data in the Data and analysis section.

Method 2: Using an illuminated object

1 Set up the equipment as shown in the diagram. Switch the lamp on behind the screen.

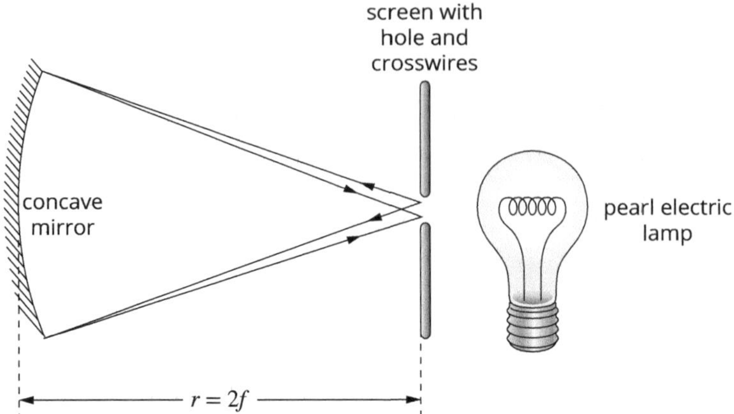

2 Move the concave mirror back and forward until a sharp image of the cross-hairs is formed on the screen close to the original cross-hairs. Measure and record the distance between mirror and screen in the Data and analysis section.

Repeat two or three times using different observers to reduce human error in determining what is a 'sharp' image. Record these in the Data and analysis section.

3 The position of the screen where there is a sharp image will correspond to the centre of curvature or radius, *r*. This is twice the focal length of the mirror. Calculate the focal length and record it in a table.

Part B: Convex lens

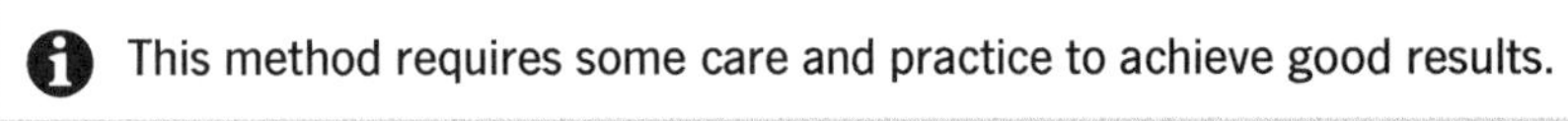

1 Place the convex lens on top of a plane mirror, reflecting side up. Use a retort stand and clamp to position the point of the pin directly above the centre of the lens, as shown below (I = image; O = object):

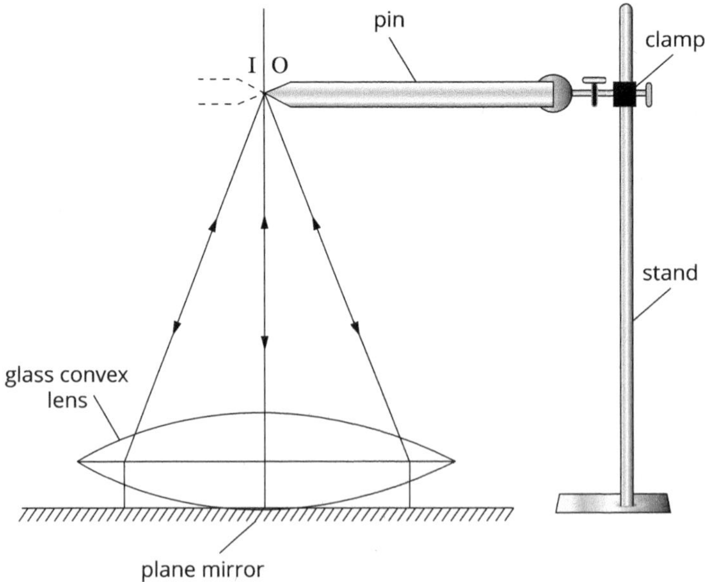

2 From directly above, look down toward the image of the pin in the mirror. Now try moving your head slightly back and forward. You will notice the image of the pin moves in relation to the pin itself. This means that the image and object do not lie in the same virtual plane (parallax error).

3 Slowly lower or raise the pin until no parallax is observed. When you move your head slightly the relative position of pin and image doesn't change. Measure the distance from the centre of the lens to the pin. This is the focal length of the lens.

4 Record your measurement in the Data and analysis section.

 ISBN 978 1 4886 1935 9

5 Move the pin out of position and repeat steps 3 or 4 three times to obtain an average. A partner could also repeat the experiment to reduce errors in personal observations, since some people are unable to observe parallax errors of this nature.

6 Average your results to determine the focal length.

DATA AND ANALYSIS

Part A: Concave mirror

Approximate focal length from Method 1: ________ cm

Average radius of curvature from Method 2: ______ cm

Focal length from Method 2: $f = \frac{r}{2} =$ ____________ cm

Part B: Convex lens

Measurement 1: ____________ cm

Measurement 2: ____________ cm

Measurement 3: ____________ cm

Measurement 4: ____________ cm

Average: ____________ cm

CONCLUSION

1 Compare the focal length determined by each method for concave mirrors. Comment on the nature of the image formed of the distant object.

2 Why should Method 2 for a concave mirror provide a more accurate determination of the focal length?

3 Comment on the result and the nature of the image formed by the convex lens.

4 Under what conditions will light rays retrace their original path when they strike the plane mirror behind the convex lens?

5 Are the light rays parallel before and after reflecting from the mirror behind the lens?

6 Why is the image of the pin inverted?

7 Only the tip of the pin is used to determine the focal length. Why?

RATING MY LEARNING	My understanding improved	Not confident ◄──► Very confident ○ ○ ○ ○ ○	I answered questions without help	Not confident ◄──► Very confident ○ ○ ○ ○ ○	I corrected my errors without help	Not confident ◄──► Very confident ○ ○ ○ ○ ○

PRACTICAL ACTIVITY 3.4

Dispersion and refraction

Suggested duration: 30+ minutes

INTRODUCTION

A prism is an optical element with flat, polished sides that refract light. When light travels from air into the prism, it is refracted or bent from its original path.

The amount of refraction depends on the angle that the incident beam of light makes with the surface of the prism, and on the ratio between the refractive indices of the air and the prism material. This relationship is described by Snell's law:

$$n_1 \sin i = n_2 \sin r$$

The refractive index of most materials also depends on the wavelength of the light. Different wavelengths refract at different angles. This phenomenon is called dispersion, and it can be used to separate a beam of white light into a spectrum of different wavelength, which we see as colours.

A ray of light refracts twice on passing through a prism—once when it enters the prism and once when it passes out of the prism. The angle through which the final ray deviates from the direction of the original incident ray is called the angle of deviation, *d*, and is a measure of the overall refraction through the prism. The diagram below shows how this can be measured.

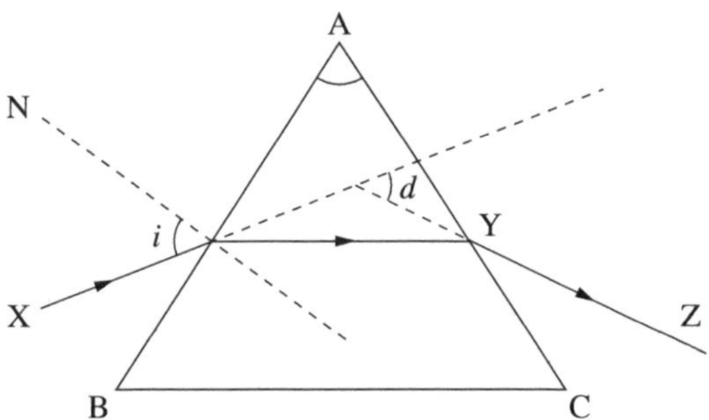

Because different wavelengths of light have different refraction indices, *d* depends on the wavelength. Shorter wavelengths (such as blue light) deviate more from their original path than colours with longer wavelengths (such as red light).

As the angle of incidence is increased, *d* decreases until it reaches a minimum value.

If the angle of incidence is increased further, *d* also increases. At the minimum angle of deviation:

- the incident angle *i* of the light ray is identical to the angle of refraction of the exit angle of the light ray
- a plane mirror placed at right angles to the exiting light ray will cause the reflection to exactly retrace its original path
- the ray of light passing through the prism is parallel to the base of the prism.

If *d* is the minimum angle of deviation, then the refractive index *n* for the particular wavelength of light for the material of the prism can be found from:

$$n = \frac{\sin\frac{1}{2}(A+d)}{\sin\frac{1}{2}A}$$

where *A* is the internal angle of the prism between the incident and refracted light (see the diagram above).

So by finding the minimum angle of deviation, the refractive index can be found for different colours from the spectrum of white light.

MATERIALS

- light box and power supply
- glass or acrylic triangular prisms
- 'magic' tape or other lightly sticking tape
- grid paper (or use the grid in the Data and analysis section)
- plane mirror
- ruler
- protractor

 ISBN 978 1 4886 1935 9

PRACTICAL ACTIVITY 3.4

PURPOSE

To determine the refractive indices of a triangular prism for red and blue light using the dispersion of white light.

PROCEDURE

1. Connect the light box to the power supply, turn it on and adjust it so that one thin ray of white light is produced. Make the ray as thin as possible.
2. Set the prism on a sheet of graph paper or the grid provided. This helps to trace the rays and measure angles. Make sure that the prism is flat and stable, and that the base of the prism is parallel to the grid. Stick the prism to the paper with tape to maintain this orientation.
3. Using the protractor, measure the value of angle A and record it in the Data and analysis section. This is the internal angle of the prism in the corner between the incident and refracted light.
4. Using the protractor, draw a normal (right angle line) from around the centre of the prism on one side of the prism. It should look similar to line N in the diagram. This will assist in measuring the angle of incidence.
5. Align the ray box to shine the ray through the prism so that it passes into the prism at the point where the normal line meets the prism.
6. Starting with the light from the ray box incident on the prism at an angle close to 90° from the normal. Slowly move the ray box so that the incident angle decreases until it is aligned with the normal, then continue to the other side of the normal until the beam is close to 90° from the normal.
7. Return the ray box to a position where the incident angle is close to 90° from the normal. Slowly move the ray box, keeping the ray of light incident on the point where the normal line meets the prism, until the red (top) side beam through the prism is parallel to the base of the prism. Use the grid lines on the paper to assist in determining that the beam is parallel.
8. Draw a line to mark the incident beam through the prism and a line for the exiting beam back through the prism. Measure the angle of minimum deviation (*d* in the diagram), and record it in the Data and analysis section for red light.
9. Repeat steps 7 and 8 at least 3 times to confirm the angle.
10. Repeat steps 7 and 8 for the blue (bottom) side of the beam. Measure and record the angle of minimum deviation *d* for blue light in the Data and analysis section.
11. Keeping the ray box in position at the minimum angle of deviation for blue light, hold the plane mirror at right angles to the emergent ray so that the mirror reflects the emergent beam back into the prism.

DATA AND ANALYSIS

PRACTICAL ACTIVITY 3.4

1 Using the angle A and the average angle of deviation of the four trials, calculate and record in the table below the values of $\frac{A}{2}$ and $\frac{A+d}{2}$.

2 Calculate the refractive index for the prism for both red and blue light and record it in the table below.

Colour	A	$\frac{A}{2}$	d Trial 1	d Trial 2	d Trial 3	d Trial 4	d, average	$\frac{A+d}{2}$	n
Red									
Blue									

CONCLUSION

1 What effect does the prism have on the beam of white light?

2 What colour of visible light is deviated the least? What colour is deviated the most? Is this what you would expect from theory?

3 What did you observe? What happened to the light as it passed through the prism as the angle of incidence was changed?

4 What is the significance of this angle?

5 What path does the light take back through the prism?

6 Find out the expected refractive index for the material of the prism used in the experiment. This is usually quoted for white light. Compare your results with the accepted value.

7 How do the experimentally determined refractive indexes of red and blue light compare with each other? Are the experimental values what is expected from theory?

8 What are the major sources of error in this experiment? How could they be reduced?

RATING MY LEARNING	My understanding improved	Not confident ⟷ Very confident ○ ○ ○ ○ ○	I answered questions without help	Not confident ⟷ Very confident 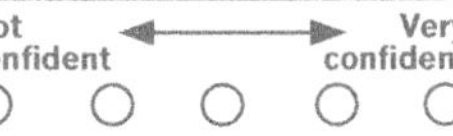	I corrected my errors without help	Not confident ⟷ Very confident ○ ○ ○ ○ ○

 ISBN 978 1 4886 1935 9

PRACTICAL ACTIVITY 3.5

Inverse square law

Suggested duration: 30 minutes

INTRODUCTION

There are actually several measures of light intensity. The simplest of these measures is irradiance, which is defined as the rate at which energy is incident upon a given surface area ($W\,m^{-2}$).

In order to understand how light intensity from a point source varies as a function of distance, consider a point light source located in space. The source emits energy at a certain rate *S*. If the energy is radiated in every direction equally, at a distance of *r* meters from the source, all of the energy generated would be passing through the surface of a sphere of radius *r*. The light intensity would be the rate of energy *S* divided by the surface area of the sphere.

$$I = \frac{S}{4\pi r^2}$$

This is known as the Inverse Square Law because it shows that the light intensity is inversely proportional to the square of the distance from the light source r^2. This can be simplified to the following relationship:

$$I \propto \frac{1}{r^2}$$

MATERIALS

- data collection system
- light sensor
- sensor extension cable
- basic optics bench
- basic optics light source
- aperture bracket

PURPOSE

To determine how the distance from a point source of light affects the intensity.

PROCEDURE

1. Set up your optics bench on a table. Mount the light source so that the line indicating the location of the point source is aligned with the 0 cm mark.
2. Mount the light sensor on the aperture bracket. Attach the aperture bracket to the holder and mount onto the optics bench.
3. Start a new activity on the data collection system. Connect the light sensor to the system using the extension cable. Configure your data collection system to monitor live data.
4. Rotate the disk on the front of the aperture bracket so the open circular aperture is in line with the opening to the light sensor. Turn on the light source.
5. Slide the sensor back and forth on the track to determine the point where the values of light intensity begin to change. This will be the minimum distance the light sensor must be away from the light source. Record the sampled data point and the distance. Record these as a point on Graph 1 in the Data and Analysis section.
6. Slide the sensor 2.0 cm further away from the source and then record another manually sampled data point and add to Graph 1. Repeat until you have collected at least 10 data points.
7. Slide the sensor 2.0 cm farther away from the source and then record another manually sampled data point.
8. Using the data collection system calculate $\frac{1}{r^2}$. Create a sketch of this data in Graph 2 in the Data and analysis section.

DATA AND ANALYSIS

Graph 1: Light intensity versus distance

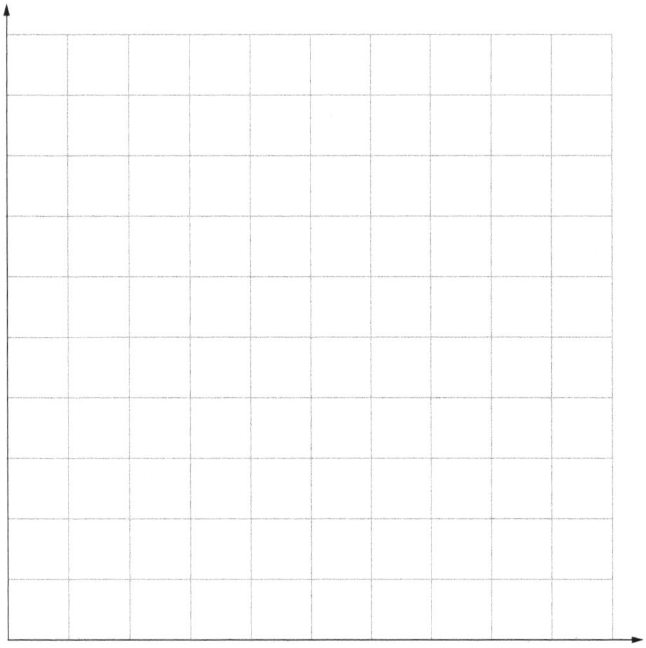

Graph 2: Light intensity versus inverse of the distance squared

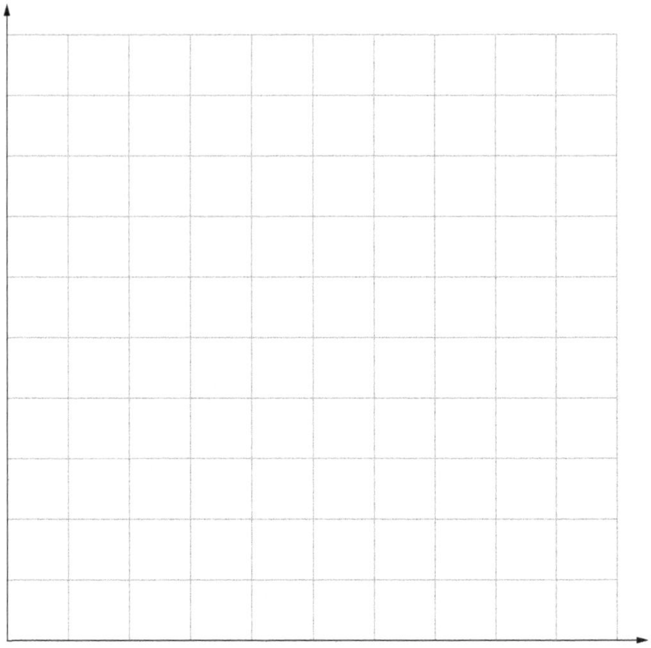

CONCLUSION

1 Compare the two graphs that you obtained. Do either of them produce a straight line? What significance does a straight-line graph represent?

RATING MY LEARNING	My understanding improved	Not confident ◄──► Very confident ○ ○ ○ ○ ○	I answered questions without help	Not confident ◄──► Very confident ○ ○ ○ ○ ○	I corrected my errors without help	Not confident ◄──► Very confident ○ ○ ○ ○ ○

 ISBN 978 1 4886 1935 9

PRACTICAL ACTIVITY 3.6

Heat capacity—the latent heat of fusion

Suggested duration: 30+ minutes

INTRODUCTION

Heat is the energy that is transferred from one substance to another as a result of a difference in their temperature. In addition to changing the temperature of a substance, heat can also break intermolecular bonds, causing the substance to change phase. When this happens, no energy goes into changing the temperature of the substance; it is all being used to alter the intermolecular bonds within the substance. The heat energy required for a substance to change phase from a solid to a liquid is given by:

$$\Delta Q = mL_{\text{fusion}}$$

where:

ΔQ is the change in energy

m is the mass of the substance changing phase

L_{fusion} is the latent heat of fusion of the substance.

The latent heat of a substance is the amount of energy required to turn it from a solid to a liquid. Like the specific heat of a substance, the latent heat of fusion depends on the type of substance and its ability to absorb heat while it changes phase.

MATERIALS

- data collection system and temperature sensor (or thermometer and stopwatch)
- 600 mL beaker
- calorimetry cup
- balance (one needed per class)
- hotplate
- stirring rod (temperature sensor can be used instead) or stir station
- 300 mL water
- 3 or 4 ice cubes
- paper towel

PURPOSE

To determine the latent heat of fusion of water.

 Keep water away from sensitive electronic equipment.

Be careful using the hotplate. Always be aware that it is on, and be conscious of any loose clothing or paper that could accidently melt or catch fire if left in contact with the hot plate.

PROCEDURE

1 Heat 300 mL of water to approximately 40°C in the beaker on the hotplate.

2 If you are using a data collection system, start a new experiment, connect the temperature sensor to the data collection system, and choose a digital display of temperature.

3 Carefully measure the mass of the calorimetry cup, and record this value in the table in the Data and analysis section.

4 Fill the calorimetry cup 3⁄4 full with hot water (approximately 40°C), then quickly (but accurately) measure the mass of the filled cup, and record this in the table in the Data and analysis section.

5 Insert the thermometer or temperature sensor into the calorimetry cup and allow the temperature to stabilise. Record the initial temperature (as accurately as possible with your equipment) in the table.

6 Dry off 3 or 4 ice cubes with the paper towel and then place the ice cubes into the cup, stirring slowly until the ice completely melts. Continue stirring for an additional minute to ensure that the water has reached equilibrium. Record the final temperature of the water in the cup in the table.

7 Remove the thermometer or temperature sensor from the calorimetry cup and use the balance to measure the total mass of the cup, initial water and melted ice. Record this value in the table.

> If you are using a stir station, add the stir bar to the calorimetry cup before weighing.

DATA AND ANALYSIS

- Calculate the initial mass of the water by subtracting the mass of the cup alone from the mass of the water and cup. Record the initial mass of water in the table.
- Calculate the mass of the ice that melted in the cup by subtracting the mass of the cup and initial water from the mass of the cup, initial water and melted ice. Record the mass of melted ice in the table.
- Calculate the change in temperature of the water in the cup by subtracting the initial temperature from the final temperature. Record the temperature change in the table.
- Calculate the change in temperature of the water that is melted from ice by subtracting the initial temperature from the final temperature. Record the temperature change in the table.
- Calculate the temperature of the melted ice.

Parameter	**Value**
Mass of calorimetry cup	kg
Mass of cup and initial water	kg
Mass of initial water	kg
Mass of cup, initial water and melted ice	kg
Mass of ice	kg
Initial temperature of water in the cup	°C
Final temperature of water in the cup	°C
Initial temperature of ice	°C
Temperature of melted ice	°C
Temperature change of water	°C
Temperature change of melted ice	°C

1 What is the sign (positive or negative) of the temperature change calculated? What does the sign signify?

2 Assuming that the specific heat capacity of water, c, is $4186\,\text{J}\,\text{kg}^{-1}\,\text{K}^{-1}$, use the values from the above table to find the heat energy transferred from the water initially in the cup.

3 Using the values calculated in the table, determine how much of the heat transferred out of the water in the cup went into warming the melted ice to the final temperature.

4 If the energy needed to warm the melted ice does not account for all the heat transferred from the water, where did the additional heat go? Using the results of your previous calculations, determine the amount of additional energy lost by the water that did not go into warming the melted ice.

 ISBN 978 1 4886 1935 9

5 Assume instead that all of the remaining energy can be attributed to the melting ice. Using the mass of the ice, calculate the latent heat of fusion of water.

6 If the theoretical value for the latent heat of fusion for water is $3.34 \times 10^5\,J\,kg^{-1}$, what is the percentage difference between the experimentally determined value and the theoretical value? Use the following formula:

$$\text{Percentage difference} = \frac{\text{theoretical value} - \text{experimental value}}{\text{theoretical value}} \times 100$$

CONCLUSION

1 Why did you dry the ice cubes before you placed them in the beaker?

2 What is the source of the heat energy that melted the ice?

3 How did the experimental value compare with the accepted theoretical value? Comment on the reliability of your answer with reference to the calculated percentage difference.

4 What assumptions were made in this experiment about the initial temperature of the ice?

5 What other assumptions were made about energy transfer in conducting this experiment?

6 Keeping these assumptions in mind, what could be done to improve the procedure and hence the experimental result?

RATING MY LEARNING	My understanding improved	Not confident ◄——► Very confident ○ ○ ○ ○ ○	I answered questions without help	Not confident ◄——► Very confident ○ ○ ○ ○ ○	I corrected my errors without help	Not confident ◄——► Very confident ○ ○ ○ ○ ○

DEPTH STUDY 3.1

Solar heating

Suggested duration: 3.5–4 hours, including data collection, analysis and report

INTRODUCTION

We often take hot water for granted: we get up out of bed in the morning, jump in the shower and the water is hot and plentiful. But there are times when the power fails, the gas is disconnected or you are on a camping trip without the luxury of instant hot water.

This depth study requires you to question how water heaters work and predict how this can be applied to constructing a working model. You will analyse data and information from practical tests and use problem-solving techniques to apply your findings to develop a better model. You will communicate your findings in a short individual or group presentation that includes your initial research, design, tests, design modification, final model and conclusions. Photos of your tests and modelling can be included to illustrate your progress.

FIGURE 1 In Victorian times it was thought that solar energy would be commonly used for heating and cooking in the future.

MATERIALS

The emphasis is on cost-effectiveness. Materials used must be general purpose and readily available. No commercial finished solar heaters can be used.

Suitable materials may include, but are not limited to:

- clean used soft drink bottles
- flexible or rigid PVC tubing and connectors
- aluminium foil
- silicon sealant
- thermometers
- stopwatches
- thermistors, data acquisition systems or other equipment readily available to you for testing.

PURPOSE

Design and construct a solar-powered water heater that maximises heating while minimising size, complexity and cost. Your design can be tested using artificial light or heat sources if there is not enough direct sunlight. The results of your testing, including heating source, change in temperature, heating time, and heat energy transferred will be used to evaluate the success of your design.

DESIGN REQUIREMENTS AND CONSTRAINTS

You can work individually or in small groups, depending on resources, time and class constraints. Check with your teacher.

No mechanical or electrical devices are permitted. The heating source must be the Sun (except that a radiant heat source such as a portable heater may be used for testing if necessary).

The water heater must be capable of heating at least 1 litre of water.

The water heater needs to heat the water at least 10°C above the initial starting temperature in less than one hour to be considered successful.

Partial prototypes may be constructed to test your design.

No commercial finished solar heaters can be used. The materials employed must be general purpose.

ISBN 978 1 4886 1935 9

SAFETY AND ACCESS

1 What safety concerns do you need to consider in this task? How will you minimise any potential risks?

QUESTIONING AND PREDICTING

2 If you were to design a water heater right now, what would be the prime considerations? Develop an inquiry question that will be the focus of your design.

3 What variables should be tested in developing your design?

4 How will you control these variables during testing?

5 Sketch two possible designs for the water heater to explain your proposals. Your sketches should include labels indicating the heating processes and proposed circulation of hot and cold water.

ISBN 978 1 4886 1935 9

6 When a material is heated, the change in temperature of the material for any given amount of energy is dependent on the material's specific heat capacity. How much energy does water require for each 1°C increase in temperature?

7 On average, 1370 W of radiant heat energy is incident on 1 m^2 of the Earth's surface from the Sun every second. Assuming 100% efficiency and no energy losses, how long would it take for the Sun to heat 1 litre of water by 1°C using 1 m^2 of heat-absorbing material?

8 Maximising efficiency while minimising heat loss will be crucial to the design. Write a hypothesis and explain how your design will consider these criteria.

9 What is the relationship between distance and the energy output from a source of light?

10 Would this influence the expected results from your design over the course of the year?

CONDUCTING YOUR INVESTIGATION

11 Using the materials available to you, run one or more tests to determine the maximum efficiency of absorption of sunlight and the potential heat loss from the proposed water container.

 ISBN 978 1 4886 1935 9

ANALYSING DATA AND INFORMATION

12 Look back at your original designs. Based on your testing, do you think the designs will work? Explain why they work or how your initial thoughts have changed and why.

PROBLEM SOLVING

13 Working in a small group, evaluate other students' designs and agree on the most effective, collaborative design. Draw the final design, including dimensions, materials and material quantities. Record the important design points and your reasons for the design.

14 Build your model water heater and run some trials under different heating conditions equivalent to the average solar radiant energy that can be expected. You may want to take photos of your model development. Using the methods available to you, measure and record the rate at which the water is heated.

COMMUNICATING

Communicate your findings using an illustrated group or individual poster. Your presentation should include your responses to the following questions along with supporting graphs and other data confirming the performance characteristics of your water heater.

15 What was the fastest time recorded by a group for a water heater to heat water 10°C above the initial temperature?

16 Look at the designs produced by other groups. What was different about their designs that would have affected their performance relative to yours? What made some designs more effective and some less effective?

17 What would you change in your design to make it more effective? Why do you think these changes would make the design more effective?

 ISBN 978 1 4886 1935 9

Multiple choice

1 As far as light is concerned, a brick exhibits which of the following optical phenomena? There may be one or more answers.

A absorption

B transmission

C reflection

D refraction

2 The following diagram shows a ray of light, X, reaching a boundary between water and glass.

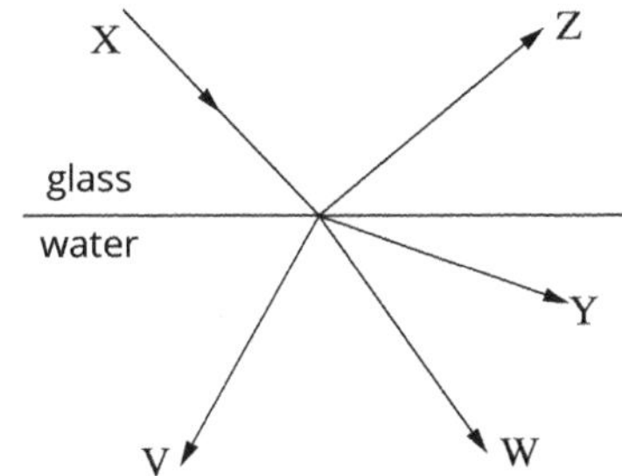

Which of the following could be paths taken by rays of light that have reached the boundary?

A W and Z

B Y and Z

C Y only

D Y and V

3 A drummer beats his drum three times every two seconds. What is the frequency of the drumming?

A 0.67 Hz

B 1.5 Hz

C 1.67 Hz

D 3.0 Hz

4 An object is placed in front of a convex mirror at a distance of twice the focal length. Which one of the following describes the image?

A upright, diminished and real

B inverted, diminished and real

C upright, diminished and virtual

D upright, enlarged and virtual

5 Which one of the following does the rate of energy transfer by conduction through a substance not depend on?

A thermal conductivity of the substance

B colour of the substance

C surface area perpendicular to the direction of heat

D temperature difference across the substance

6 A mosquito is resting on the surface of a pond when a ripple approaches.

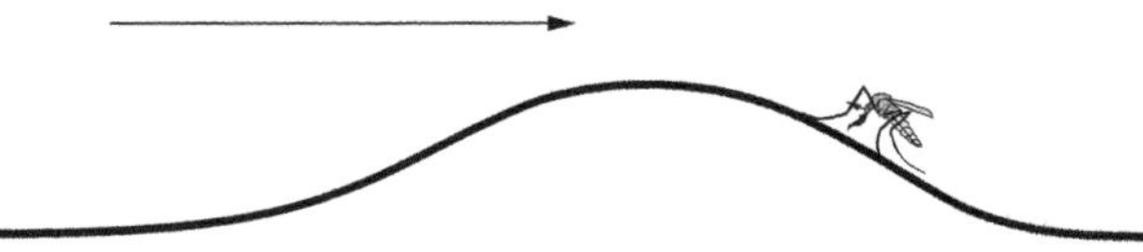

At the time shown, what is the motion of the mosquito?

A It is moving to the right.

B It is moving to the left.

C It is moving up.

D It is moving down.

E It is stationary.

Short answer

7 The graph below shows a 25 Hz sound wave recorded by a microphone. The horizontal scale is in metres, while the vertical scale consists of arbitrary pressure units.

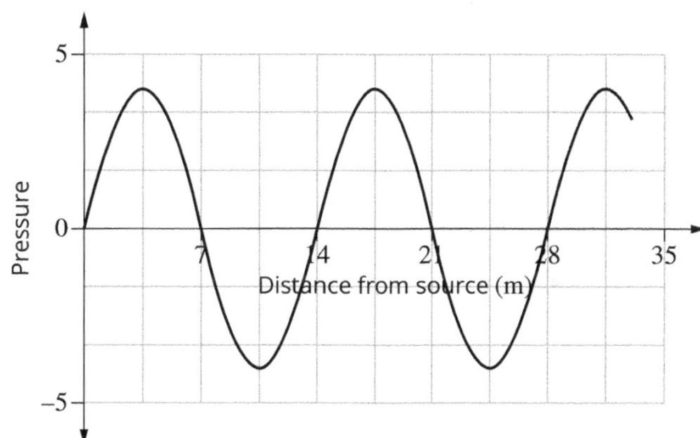

a What is the amplitude of this sound wave?

b What is the wavelength?

c Calculate the speed of the sound.

8 An object 3.0 cm high is placed 40.0 cm from a concave mirror whose focal length is 30.0 cm.

a Calculate the location and height of the image, and state whether it is real or virtual.

b Repeat for the same object placed 40 cm from a convex lens.

9 When a police car was 100 m away from a stationary observer, its siren emitted sound that was recorded as having a power of 2.5×10^{-10} W m^{-2} and a frequency of 420 Hz. As it approached the observer, the recorded power increased to 4.0×10^{-9} W m^{-2}.

At the point when the police car was exactly alongside the observer, the frequency of the sound was measured as 400 Hz, and after it passed the observer its frequency was recorded as 380 Hz.

a How far away was the police car when the 4.0×10^{-9} W m^{-2} sound reading was made?

b Assuming the speed of sound is 340 m s^{-1}, how fast was the police car travelling?

10 A 200 g sample of naphthalene is heated carefully in a closed glass vessel in which all the fumes are safely contained and the energy inputs can be accurately measured. The heating curve below was produced as a result of the measurements.

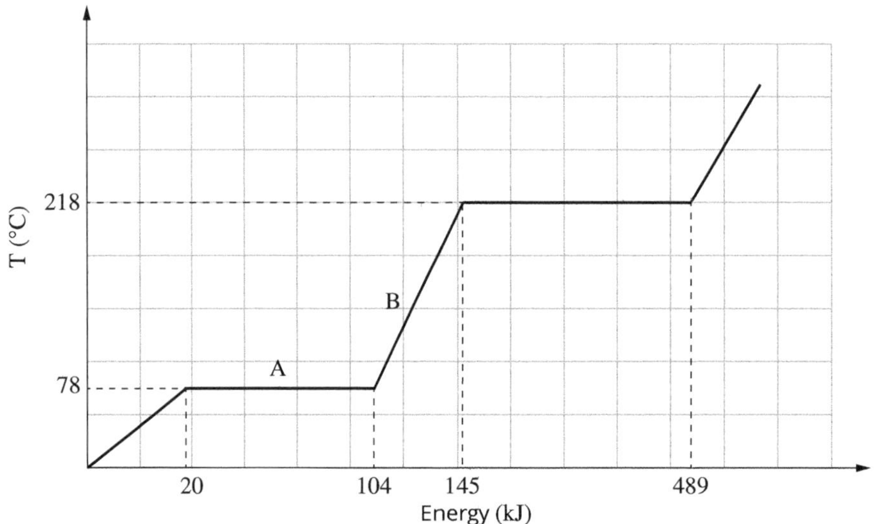

a What states of matter would be found in the heating vessel in sections A and B?

Use the information on the graph to calculate the following.

b The specific heat capacity of solid naphthalene.

c The specific heat capacity of liquid naphthalene.

d The latent heat of fusion of naphthalene.

e The latent heat of vaporisation of naphthalene.

 ISBN 978 1 4886 1935 9

Extended response

11 Isaac Newton famously showed that white light is composed of many different colours by passing it through a prism. This phenomenon is known as dispersion and is caused by different colours having slightly different refractive indices in most transparent materials. For example, in crown glass, violet light ($\lambda = 400\,\text{nm}$) has a refractive index of 1.53 whereas red light ($\lambda = 700\,\text{nm}$) has a refractive index of 1.51.

a The focal length of a lens depends on the shape of its surface and its refractive index. Explain what effect these two factors have on the focal length.

b The diagram below depicts an ideal image formation of a star on a white screen.

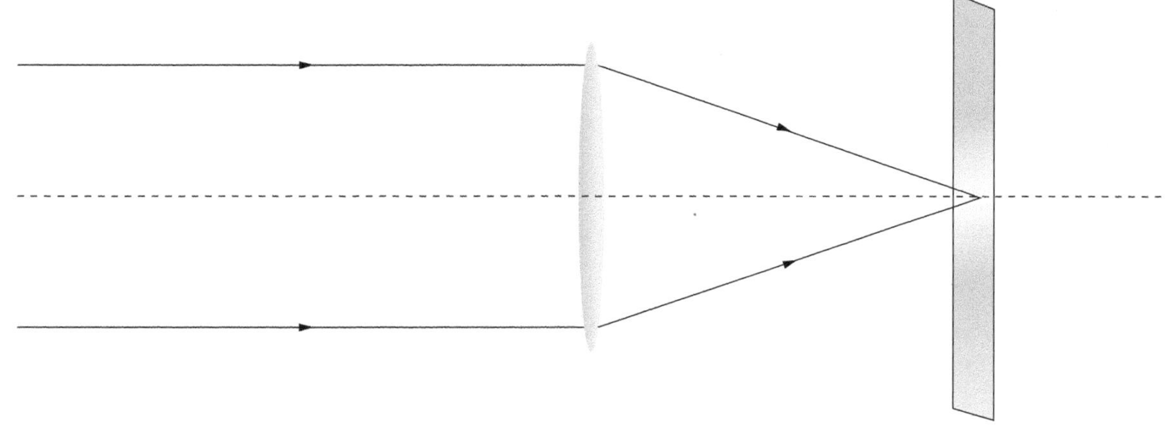

Label the diagram, indicating the position of the star and its image, together with the lens, screen and the focal length of the lens.

c If the rays depicted have a wavelength of 450 nm, draw on the diagram the paths taken by red and violet light. (You may exaggerate the bending of the rays for clarity.)

d What effect will this have on the star's image?

e What does this mean for the measurement of the focal length of lenses?

12 A series of wave crests, λ_A apart, travels from section A of a water tank into section B. As they enter section B, the distance between the wave crests becomes λ_B. The incident and refracted wave crests make angles of θ_A and θ_B respectively with the boundary. A magnified section of the boundary showing two successive wave crests is also shown.

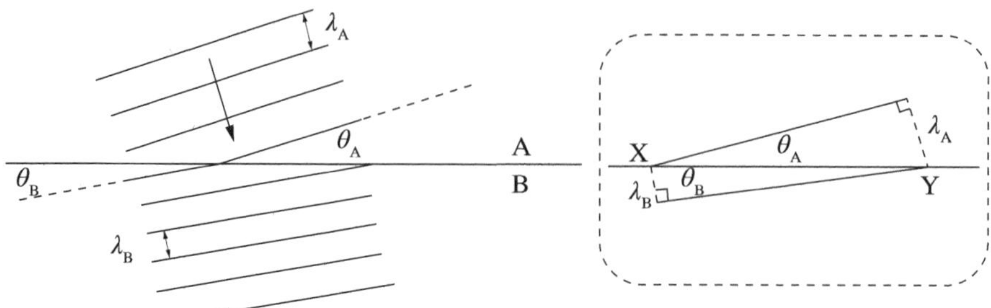

a In the diagram, there is a pair of right-angled triangles which share the line XY as the hypotenuse. Write expressions for $\sin\theta_A$ and $\sin\theta_B$.

b Now write an expression for $\frac{\sin\theta_A}{\sin\theta_B}$.

c Compare the frequencies of the waves in the two sections.

d By using the wave equation, extend your expression in part **b** to include the wave speeds v_A and v_B.

e Which section of the water tank is deeper, A or B?

f Christiaan Huygens proposed a wave model for light. Describe how the wave model explains refraction and how the expression you have found relates to Snell's law.

 ISBN 978 1 4886 1935 9

MODULE 4 Electricity and magnetism

Outcomes

By the end of this module you will be able to:

- develop and evaluate questions and hypotheses for scientific investigation PH11-1
- analyse and evaluate primary and secondary data and information PH11-5
- communicate scientific understanding using suitable language and terminology for a specific audience or purpose PH11-7
- explain and quantitatively analyse electric fields, circuitry and magnetism PH11-11

Content

ELECTROSTATICS

INQUIRY QUESTION **How do charged objects interact with other charged objects and with neutral objects?**

By the end of this module you will be able to:

- conduct investigations to describe and analyse qualitatively and quantitatively: CCT ICT
  - processes by which objects become electrically charged (ACSPH002)
  - the forces produced by other objects as a result of their interactions with charged objects (ACSPH103)
  - variables that affect electrostatic forces between those objects (ACSPH103)
- using the electric field lines representation, model qualitatively the direction and strength of electric fields produced by:
  - simple point charges
  - pairs of charges
  - dipoles
  - parallel charged plates ICT
- apply the electric field model to account for and quantitatively analyse interactions between charged objects using: ICT N
  - $\vec{E} = \frac{\vec{F}}{q}$ (ACSPH103, ACSPH104)
  - $\vec{E} = -\frac{V}{\vec{d}}$
  - $\vec{F} = \frac{1}{4\pi\varepsilon_0} \times \frac{q_1 q_2}{r^2}$ (ACSPH102)
- analyse the effects of a moving charge in an electric field, in order to relate potential energy, work and equipotential lines, by applying: (ACSPH105)
  - $V = \frac{\Delta U}{q}$, where U is potential energy and q is the charge

ELECTRIC CIRCUITS

INQUIRY QUESTION **How do the processes of the transfer and the transformation of energy occur in electric circuits?**

Module 4 • Electricity and magnetism

By the end of this module you will be able to:

- investigate the flow of electric current in metals and apply models to represent current, including:
  - $I = \frac{q}{t}$ (ACSPH038) CCT ICT N
- investigate quantitatively the current–voltage relationships in ohmic and non-ohmic resistors to explore the usefulness and limitations of Ohm's Law using:
  - $V = \frac{W}{q}$
  - $R = \frac{V}{I}$ (ACSPH003, ACSPH041, ACSPH043) ICT N
- investigate quantitatively and analyse the rate of conversion of electrical energy in components of electric circuits, including the production of heat and light, by applying $P = VI$ and $E = Pt$ and variations that involve Ohm's Law (ACSPH042) ICT N
- investigate qualitatively and quantitatively series and parallel circuits to relate the flow of current through the individual components, the potential differences across those components and the rate of energy conversion by the components to the laws of conservation of charge and energy, by deriving the following relationships: (ACSPH038, ACSPH039, ACSPH044) ICT N
  - $\Sigma I = 0$ (Kirchhoff's current law—conservation of charge)
  - $\Sigma V = 0$ (Kirchhoff's voltage law—conservation of energy)
  - $R_{\text{series}} = R_1 + R_2 + ... + R_n$
  - $\frac{1}{R_{\text{parallel}}} = \frac{1}{R_1} + \frac{1}{R_2} + ... + \frac{1}{R_n}$
- investigate quantitatively the application of the law of conservation of energy to the heating effects of electric currents, including the application of $P = VI$ and variations of this involving Ohm's law (ACSPH043) CCT N

MAGNETISM

INQUIRY QUESTION How do magnetised and magnetic objects interact?

By the end of this module you will be able to:

- investigate and describe qualitatively the force produced between magnetised and magnetic materials in the context of ferromagnetic materials (ACSPH079)
- use magnetic field lines to model qualitatively the direction and strength of magnetic fields produced by magnets, current-carrying wires and solenoids and relate these fields to their effect on magnetic materials that are placed within them (ACSPH083) ICT
- conduct investigations into and describe quantitatively the magnetic fields produced by wires and solenoids, including: (ACSPH106, ACSPH107)
  - $B = \frac{\mu_0 I}{2\pi r}$ ICT N
  - $B = \frac{\mu_0 NI}{L}$ ICT N
- investigate and explain the process by which ferromagnetic materials become magnetised (ACSPH083)
- apply models to represent qualitatively and describe quantitatively the features of magnetic fields ICT N

Key knowledge

Electrostatics

ELECTRIC CHARGE

Matter consists of atoms. Atoms have a nucleus consisting of positively charged particles called protons and neutral particles called neutrons. (One exception is hydrogen, which in its normal state has only one proton and no neutrons.) Surrounding the nucleus are negatively charged electrons, which are attracted to the positively charged nucleus. This attraction is one of the fundamental forces of nature. In electrostatic forces, like charges repel and unlike charges attract.

An atom is said to be electrically neutral if the number of electrons balances the number of protons. As electrons in an atom are in a cloud surrounding the nucleus, they can be separated from the atom with relative ease, whereas the protons are held tightly in the nucleus. When an atom loses an electron it develops a positive net charge; when it gains an electron, it develops a negative net charge.

The net charge is often referred to as the **elementary charge.** The letter q is used to represent the amount of charge. The SI unit of charge is the coulomb (C), where one coulomb is the charge on 6.25×10^{18} electrons or protons. Therefore, the charge of a proton (q_p) is equal to 1.6×10^{-19} C and the charge of an electron (q_e) is -1.6×10^{-19} C.

The total charge can then be calculated by multiplying the number of elementary particles by their specific charge. For example, the total charge carried by 2 billion electrons would be:

$$\begin{aligned} q &= n_e \times q_e \\ &= 2 \times 10^{9} \times -1.6 \times 10^{-19} \\ &= -3.2 \times 10^{-10}\,\text{C} \end{aligned}$$

The behaviour of an object that has been charged depends upon the material it consists of; it will either be conductive or non-conductive (insulators). Electrons move easily through conductors, but not through insulators. This is because the electrons in materials that are good conductors are weakly attracted to the nucleus, and electrons in insulators are more strongly attracted to the nucleus. Most metals are more conductive than non-metals, which are mostly insulative.

All electrically charged objects produce an electrostatic force on other objects that causes like charges to repel and opposite charges to attract. The magnitude of the electrostatic force can be determined using Coulomb's law, which is discussed later.

ELECTRIC FIELDS

An electric field is a region of space around a charged object in which another charged object will experience a force. This force is either attractive or repulsive, depending on the charges of the two objects. Although this is a non-contact force like a gravitational field, gravity only applies an attractive force. An electric field is a vector quantity because it has both a magnitude and a direction.

Electric fields are represented by field lines, which are a convenient way of visualising the field. When drawing electric field lines (in two dimensions) around a charged object there are a few rules that need to be followed:

- The field lines point in the direction of the force that a positive charge within the field would experience.
- A positive charge experiences a force in the direction of the electric field, and a negative charge experiences a force in the opposite direction to the field.
- The spacing between the field lines indicates the strength of the field. The closer together the lines are, the stronger the field.
- Around point charges, which can be thought of as small charged spheres with no spatial extent, the electric field radiates in all directions (three dimensionally).
- Between two oppositely charged parallel plates, the field lines are parallel and therefore the field has a uniform strength.
- When charged particles are in an electric field, they accelerate in the direction of the force acting on them.

These rules for electric field lines have been established to communicate the greatest amount of information about the nature of the electric field surrounding a charged object. Figure 4.1 illustrates grass seeds suspended in oil aligning themselves with an electric field. The electric field lines in each photo are shown in the diagrams on the right.

Electric field strength can be expressed as:

$\vec{E} = \frac{\vec{F}}{q}$ where $\vec{F}$ = the force on the charged particle (in N), q = the charge of the object experiencing the force (in C) and $\vec{E}$ = the strength of the electric field (in N C^{-1}).

Remember that a positive or negative sign attached to a vector will indicate direction. Also, that a positive charge in an electric field will experience a force in the same direction as the electric field, so a negative charge will experience a force in the opposite direction to the electric field.

You can also rearrange this formula, so that $\vec{F} = q\vec{E}$. Force can be related to the acceleration of a particle using the equation

$$\vec{F} = m\vec{a}$$

where:

m is the mass of the accelerating particle (in kg)

$\vec{a}$ is the acceleration (in m s^{-2}).

The electric field strength can also be expressed as

$$\vec{E} = -\frac{V}{\vec{d}}$$

where:

V is the electrical potential or voltage (in V)

$\vec{E}$ is the electrical field strength (in V m^{-1})

$\vec{d}$ is the displacement or distance between points, parallel to the field (in m).

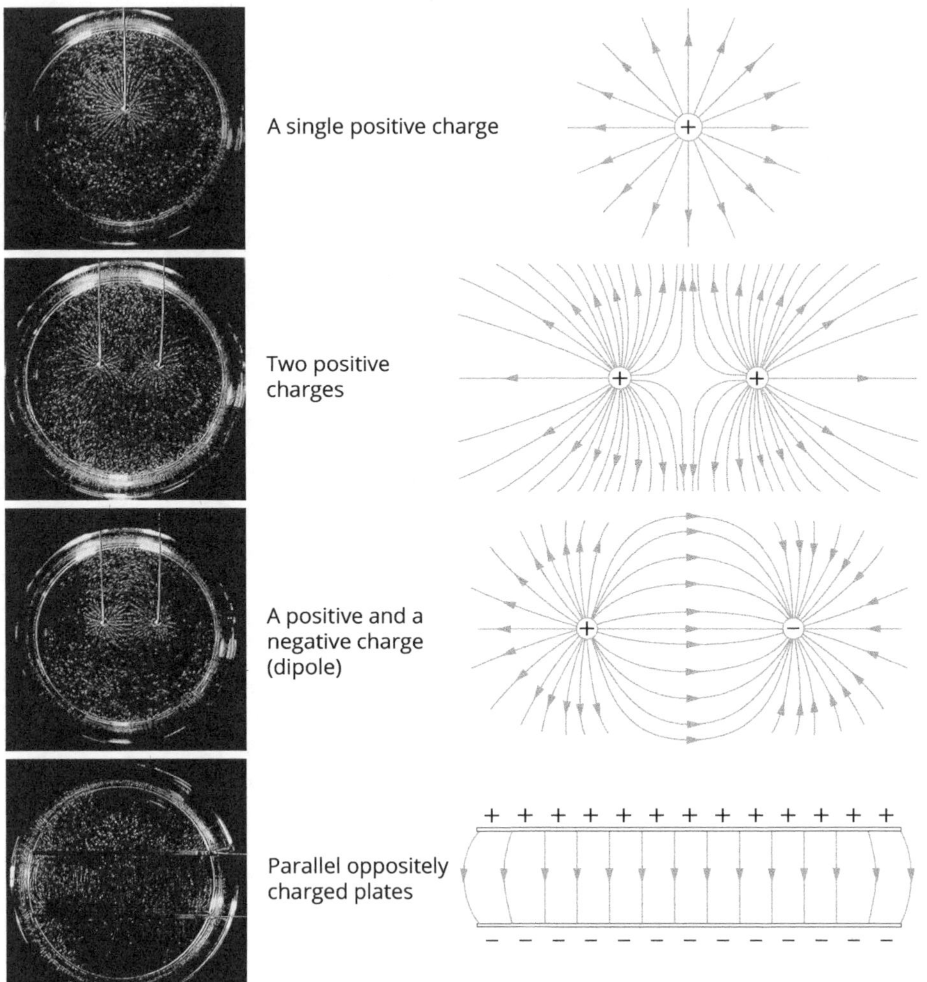

FIGURE 4.1 Electric field diagrams

Electrical potential (V) can be determined using the equation

$$V = \frac{\Delta U}{q}$$

where ΔU is the difference in electric potential energy (in J).

This is equivalent to the work done per unit charge. This is discussed further in the section 'Energy in electric circuits'.

There is an electrical potential in the space between two oppositely charged plates, as shown in Figure 4.2.

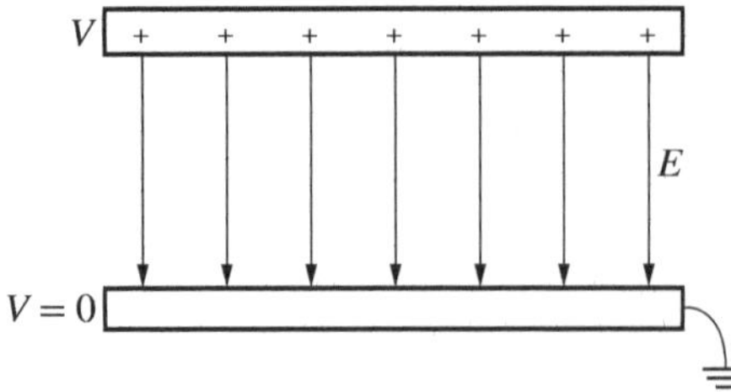

FIGURE 4.2 Electric potential between two oppositely charged plates

Electrical potential energy is a form of energy that is stored in an electric field. Work is done to move electric charges of the same polarity together, as the force of mutual repulsion is overcome.

When a charged object is moved against the direction it would naturally move in an electric field, then work is done on the field. When a charged object is moved in the direction it would naturally tend to move in an electric field, then the field does work on the particle. If a charged particle moves perpendicular to the field then no work is done. The work done on or by an electric field can be calculated using the equations $W = qV$ or $W = qEd$.

COULOMB'S LAW

Coulomb's law gives the force between two point charges. The force is along the line connecting the charges, and is attractive if the charges are opposite, and repulsive for like charges.

Coulomb's law for the force between two charges q_1 and q_2 (in C) separated by a distance r (in m) is given by the following equation:

$$\vec{F} = \frac{1}{4\pi\varepsilon_0}\frac{q_1 q_2}{r^2}$$

where ε_0 is a constant with a value of $8.8542 \times 10^{-12}\,C^2 N^{-1} m^{-2}$.

The expression $\frac{1}{4\pi\varepsilon_0}$ at the front of the Coulomb's law equation can be simplified to the value of k, called Coulomb's constant, which has a value of approximately $9.0 \times 10^9\,N\,m^2\,C^{-2}$. So:

$$\vec{F} = k\frac{q_1 q_2}{r^2}$$

If there are multiple point charges, the forces add by superposition (the head-to-tail vector method).

 ISBN 978 1 4886 1935 9

The direction of the force will either describe repulsion (two like charges), or attraction (two opposite charges).

It is possible to combine Coulomb's law with the strength of the electric field to find the magnitude:

$$E = k\frac{q}{r^2}$$

where $k = 9.0 \times 10^9\,\mathrm{N\,m^2\,C^{-2}}$.

Electric circuits

ELECTRIC CURRENT AND CIRCUITS

The movement of an electric charge in a wire is called an electric current. Electrons move easily through conductors, so wiring in electric circuits needs to consist of a good conductor such as a metal. As an electron collides with the neighbouring electron, it transfers energy. This energy can be converted into many other forms, such as heat, light and sound energy. For the electrons to flow, a path or electric circuit is required. Current will flow in a circuit only when the circuit forms a continuous (closed) loop from one terminal of a power supply to the other terminal. In the loop, usually different electrical components such as a light bulb or switch are added.

Circuit diagrams are used to clearly show how the components of an electric circuit are connected. Figure 4.3a is a cutaway diagram of a torch, showing how a battery, light bulb and switch are physically connected by a conductive material in a torch. In contrast, Figure 4.3b is a much simpler circuit diagram of the torch. This also shows how the components are connected, using standard symbols for the components.

When an electric current flows, electrons all around the circuit move towards the positive terminal at the same time. This is called electron flow. Conventionally, current in a circuit is said to flow from the positive terminal to the negative terminal.

Current (symbol I) is defined as the amount of charge q that passes through a point in a conducting wire per second. It has the unit amperes or amps (A), which are equivalent to coulombs per second. If the number of electrons (n_e) that flow through a particular point in the circuit is known, then this can be multiplied by the charge of one electron ($q_e = -1.6 \times 10^{-19}\,\mathrm{C}$) and divided by the time to find the current. The equation for this is:

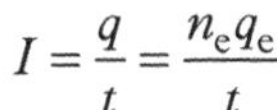

$$I = \frac{q}{t} = \frac{n_e q_e}{t}$$

Current is measured with an ammeter connected along the same path as the current flowing (that is, in series) within the circuit. Nowadays the job of an ammeter is often done with another, more versatile instrument called a multimeter, which can measure more than just current.

ENERGY IN ELECTRIC CIRCUITS

Electrons must have energy to move around a circuit. This energy can be provided by a cell or battery. The chemical energy within the cell is then transformed into electric potential energy.

Electric potential difference is a measure of the difference in electric potential energy available per unit charge.

Potential difference can be defined as the work done to move a charge against an electric field between two points, using the equation:

$$W = Vq \text{ or } V = \frac{W}{q}$$

Voltage (potential difference) is usually measured by a device called a voltmeter. Unlike an ammeter, which measures the current passing through a wire, a voltmeter measures the change in voltage (the potential difference) as current passes through a component or section of the circuit. One wire of the voltmeter is connected to the circuit before the component and the other wire is connected to the circuit after the component. This is called connecting the voltmeter 'in parallel'.

When electrical current passes through a light bulb, its electrical energy is converted by the bulb into light and heat. The greater the current flow, the brighter the bulb will be. The rate at which energy is used or consumed (converted into light and heat energy) over a certain time is known as its electrical power. Power (P) is the rate at which energy is transformed in a circuit component, and is measured in watts (W). It is defined and quantified by the relationship:

$$P = \frac{E}{t} = VI$$

where P is the power (in W).

1 watt (W) is equivalent to $1\,\mathrm{J\,s^{-1}}$.

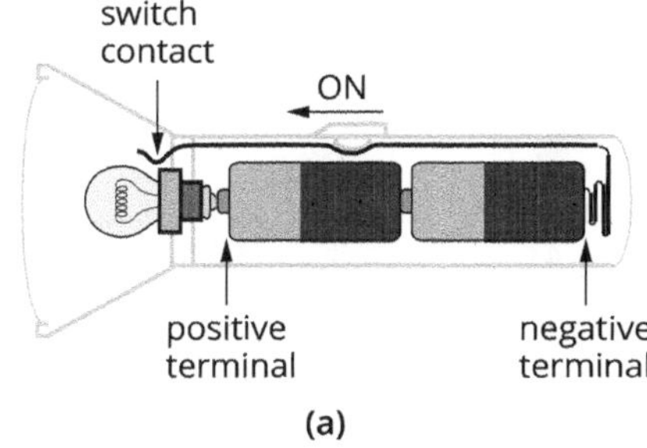

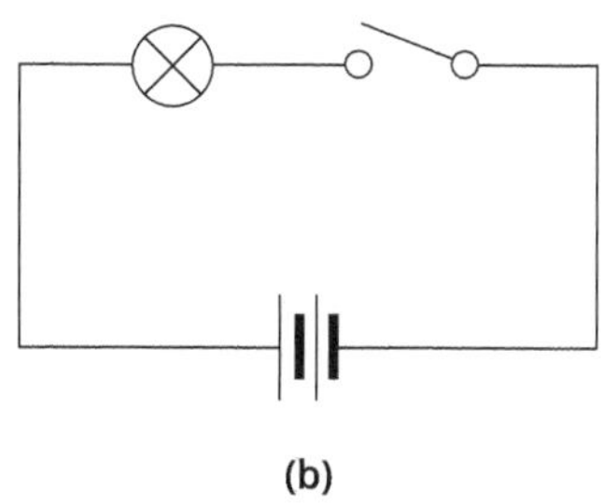

FIGURE 4.3 (a) A cutaway diagram of the physical structure of a torch and (b) the circuit diagram for the torch

RESISTANCE

Resistance is a measure of how hard it is for current to flow through a material. Resistance is measured in ohms (Ω). The resistance of a material depends on its length, cross-sectional area and temperature. If the material has a longer length, this means a greater likelihood of collisions with other ions within the conductor, making it harder for the current to flow. This is also why cross-sectional area and temperature affect the resistance. With a large cross-sectional area, there is less chance of collisions, and the opposite with temperature—the higher the temperature means the higher the vibration of the ions, making the chances of electrons colliding higher.

Ohm's law describes the relationship between current, potential difference (voltage) and resistance:

$$V = IR$$

where:

I is the current (in A)

R is resistance (in Ω).

For example, if a potential difference of 12V is applied across a piece of wire, and a 4A current flows, so the resistance of the wire is:

$$R = \frac{V}{I} = \frac{12}{4}$$
$$= 3\,\Omega$$

Not all conductors obey Ohm's law; those that do are known as **ohmic** conductors. Ohmic conductors have a constant resistance. The resistance of non-ohmic conductors varies for different potential differences. An example of a non-ohmic conductor is a light bulb. To investigate whether a material is ohmic or non-ohmic, an I–V graph can be plotted. The graph for an ohmic conductor (or resistor) is a straight line, and the resistance can be found from the gradient of the graph, as shown in Figure 4.4a. The I–V graph for a non-ohmic conductor is not a straight line, as shown in Figure 4.4b.

SERIES AND PARALLEL CIRCUITS

When a circuit contains more than one resistor, Ohm's law alone is not sufficient to predict the current flowing through and the potential difference across each resistor. To do this, the circuit can be broken up into sections depending on whether the circuit components are in series or parallel.

A series circuit has the electrical components connected one after another, so there is only one path for current to flow. For example, if one light bulb fails or one switch is opened, the whole circuit is broken. A parallel circuit contains more than one path for current to flow, so if one light bulb goes out or one switch is opened, the current can continue to flow along another path. Parallel circuits allow individual components to be switched on and off independently.

When resistors are connected in series, the following conditions occur:

- The current through each resistor is the same.
- The sum of the potential differences is equal to the potential difference provided to the circuit. This is also known as Kirchhoff's voltage law.
- The equivalent effective resistance is equal to the sum of the individual resistances; that is:

$$R_{series} = R_1 + R_2 + \ldots + R_n.$$

Equivalent resistances are used in circuit analysis to simplify a complicated circuit diagram so that current and potential difference can be determined. This is also the same in parallel circuits.

When resistors are connected in parallel, the following conditions occur:

- The voltage across each resistor is the same.
- The current is shared between the resistors. This is known as Kirchhoff's current law.
- The equivalent effective resistance is given by the equation $\frac{1}{R_{parallel}} = \frac{1}{R_1} + \frac{1}{R_2} + \ldots + \frac{1}{R_n}$.

Analysing a complex circuit may involve the calculation of both equivalent series and equivalent parallel resistances.

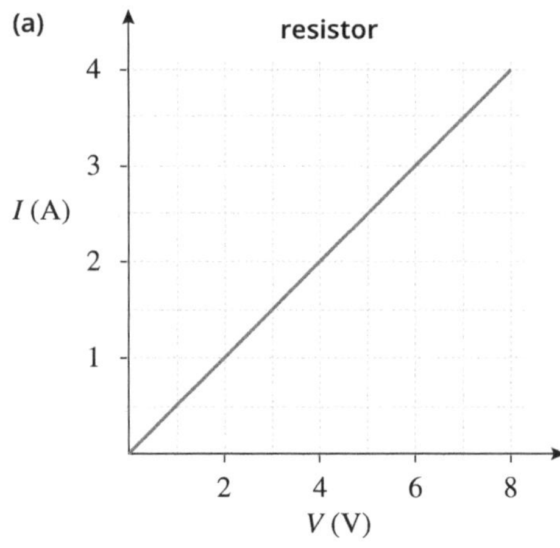

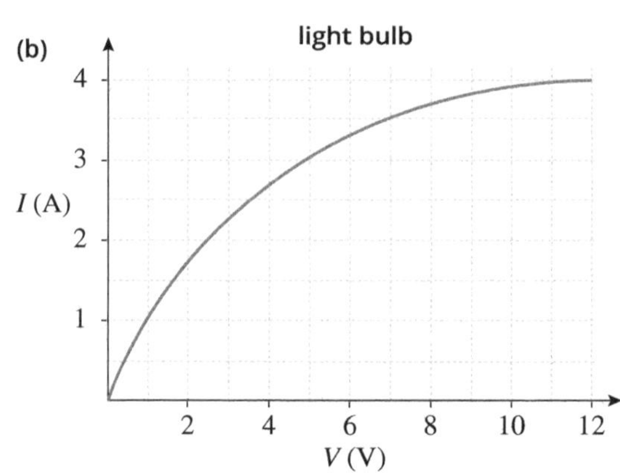

FIGURE 4.4 (a) An ohmic conductor and (b) a non-ohmic conductor

ISBN 978 1 4886 1935 9

In Figure 4.5, resistors R_2 and R_3 are in parallel, so the current is split between them. This means that the equivalent effective resistance of this pair of resistors is found as follows:

$$\frac{1}{R_{\text{parallel}}} = \frac{1}{30} + \frac{1}{40}$$
$$= \frac{7}{120}$$
$$R_{\text{parallel}} = \frac{120}{7}$$
$$= 17.1\ \Omega$$

This component, R_{parallel}, is then in series with the resistors R_1 and R_4, with the same current running through each component, so that the total resistance of the circuit is $R_{\text{series}} = 10 + 17.1 + 20 = 47.1\ \Omega$.

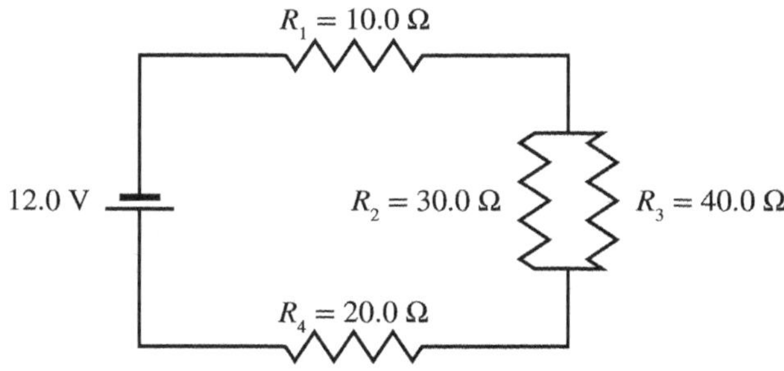

FIGURE 4.5 A series and parallel circuit

A combination of resistors draw different amounts of power depending on whether the resistors are wired in series or parallel. A parallel circuit generally draws more power than a series circuit using the same resistors. This is because resistors in parallel draw more current, as the current always prefers the path of least resistance.

Magnetism

MAGNETIC MATERIALS

Since ancient times, certain substances, called magnets, have been known to have the property of attracting iron and some other metals. This attractive property is called magnetism. Magnetism, like electrostatic force, results in a field that attracts or repels other magnetic materials. As with electrostatic force, like magnetic poles repel and unlike magnetic poles attract. Magnetic poles exist only as dipoles, having both north and south poles. A single magnetic pole (monopole) is not known to exist.

Every material experiences magnetism to some extent, some more strongly than others. **Ferromagnetic materials** such as iron, cobalt and nickel are very easily magnetised. For ferromagnetic materials to become magnets, the majority of the magnetic domains in the material must be aligned. A magnetic domain is a region in the material where the magnetic field is aligned. The magnetic field in separate magnetic domains that point in different directions is an unmagnetised material, shown in Figure 4.6a. The magnetic field in separate magnetic domains that point in the same direction is a magnetised material, shown in Figure 4.6b.

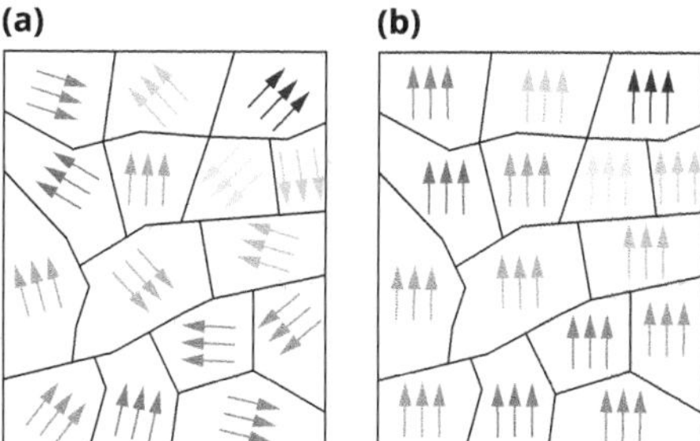

FIGURE 4.6 (a) Unmagnetised and (b) magnetised materials

MAGNETIC FIELDS

Magnetic field lines can be described by imagining a tiny compass placed at nearby points. The direction of a magnetic field is from the magnetic north pole to the magnetic south pole, which is the direction a compass would point. The denser (closer) lines indicate a stronger magnetic field. Figure 4.7 illustrates a magnet with the magnetic field lines drawn from the magnetic north pole to the magnetic south pole, and denser regions at the two pole ends. The magnetic field ($\vec{B}$) is a vector quantity because it has both a magnitude and a direction. The strength, or vector magnitude, of the magnetic field is denoted by the variable $\vec{B}$ and the unit is the tesla (T).

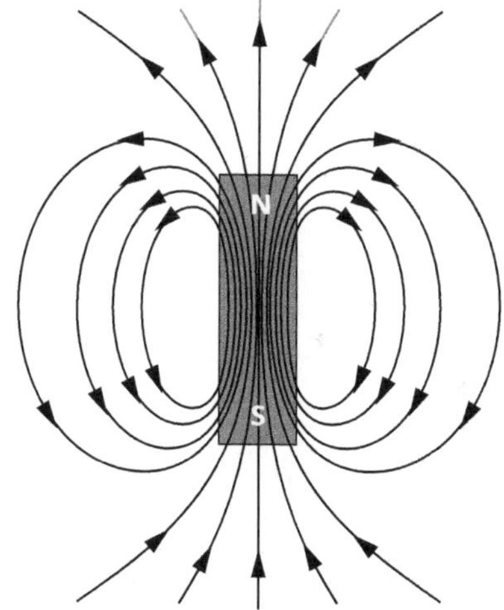

FIGURE 4.7 Magnetic field lines for a magnet

Magnetic field lines can either be uniform or non-uniform, depending on the shape of the magnet. A uniform distribution of field lines represents a uniform magnetic field. A non-uniform field, such as that around a non-circular coil, is shown by variations in the separation of the field lines.

Figure 4.7 illustrates a static (stationary) magnetic field. A magnetic field is static if it has a fixed strength and position. Where the magnetic field is changing, such as that associated with an alternating current direction, the magnetic field will also be changing.

Many electrical devices such as motors, generators, transformers and electromagnets require a magnetic field, which is usually produced by an electric current.

An electromagnet is a device in which an electric current is used to create or increase a magnetic field.

Charged particles that are moving, such as the electrons in a conductive material, create a magnetic field. This field is circular around the current-carrying conductor. The direction of the field can be determined using the **right-hand grip rule** if the direction of the current is known. An example of this is shown in Figure 4.8a, where the magnetic field lines are shown by iron filings around a current-carrying wire. Figure 4.8b indicates the right-hand grip rule. If the thumb points in the direction of the conventional current (i.e. from positive to negative), the fingers curl around the conductor in the direction of the magnetic field.

This rule can also be used to find the direction of conventional current if the direction of the magnetic field is known.

'Conventional current' is from positive to negative, because it was originally thought that this is how current flows. We now know that the current actually flows in the opposite direction. This actual flow is called 'electron current'.

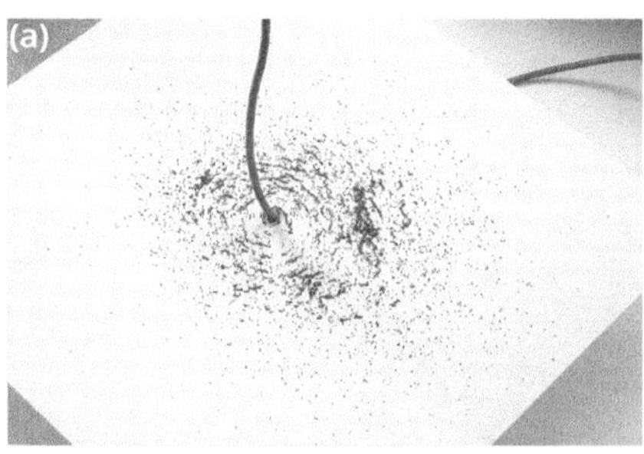

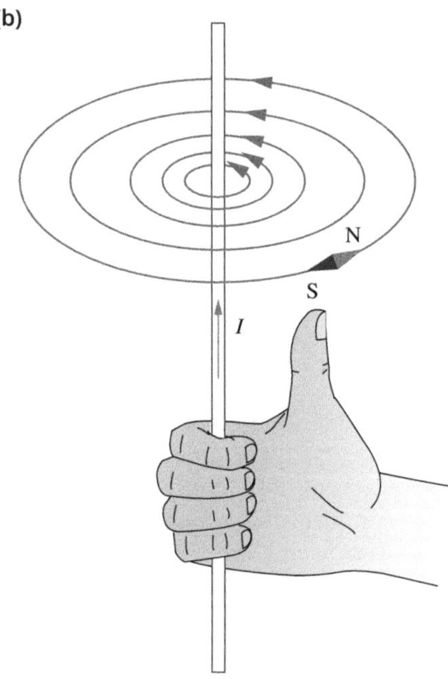

FIGURE 4.8 (a) A current-carrying conductor and (b) the right-hand grip rule.

A wire bent into a single circular loop can be thought of as a series of small straight segments that create a large magnetic field. A device commonly known as a **solenoid** demonstrates this (Figure 4.9). The field around the solenoid is like the field around a normal bar magnet; it contains an effective north end on the left and a south end on the right. The compass points in the direction of the field lines.

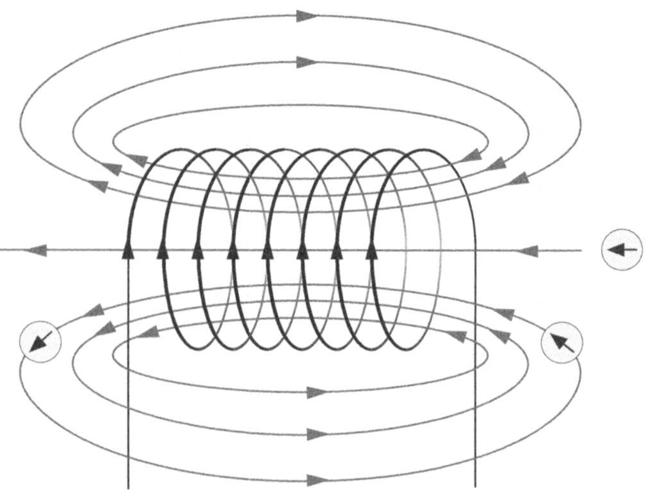

FIGURE 4.9 Magnetic field lines of a solenoid

More complex fields can be determined by applying the right-hand grip rule to the loops or coils making up the current-carrying conductor in a solenoid.

The strength of a magnet can be increased by wrapping a current-carrying wire around the magnet to create an electromagnet. The atoms in an unmagnetised ferromagnetic material such as iron point in random directions, so that the individual magnetic fields tend to cancel each other out. However, the magnetic field produced by wires wrapped around an iron core force the atoms within the core to point in one direction. Their individual magnetic fields add together, creating a stronger magnetic field.

CALCULATING MAGNETIC FIELDS

A current-carrying wire produces a magnetic field in the form of concentric circles surrounding the wire. The magnitude of that magnetic field is defined by Ampere's law. Ampere's law states that the sum of all the magnetic field elements that make up the circle surrounding the wire (with radius r) is the product of the current in the wire (I) and the permeability of free space (μ_0), which is approximately equal to $1.257 \times 10^{-6}\,\mathrm{N\,A^{-2}}$. This is represented in the following equation:

$$\vec{B} = \frac{\mu_0 I}{2\pi r}$$

The magnetic field at a point well inside a solenoid is uniform and independent of the length or diameter of the solenoid. Instead, it is dependent upon the number of turns of the coil (N) per unit length (L) of the solenoid. This equation can be modified to:

$$\vec{B} = \frac{\mu_0 N I}{L}$$

The more tightly wound the wire around a solenoid, the greater the magnetic field. Solenoids can be used to generate nearly uniform magnetic fields that are similar to the magnetic field of a bar magnet.

 ISBN 978 1 4886 1935 9

WORKSHEET 4.1

Knowledge review—electricity and magnetism

1 Match the symbols on the right with the names of the circuit components on the left.

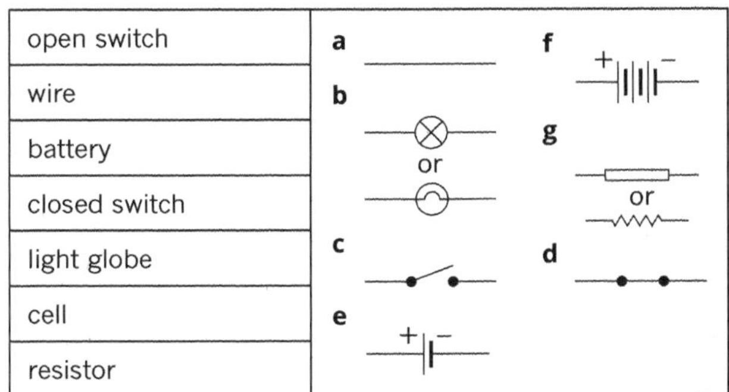

Component	Symbols
open switch	a
wire	b
battery	c
closed switch	d
light globe	e
cell	f
resistor	g

2 State the SI unit and its symbol for each of the following quantities.

Quantity	**SI unit**	**Symbol**
charge		
current		
magnetic field strength		
potential difference		
power		
resistance		

3 Consider the following circuit containing a cell and two identical globes, and fill in the blanks below.

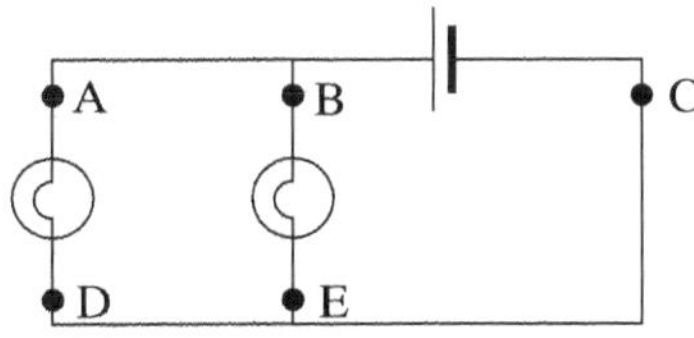

a This is a ____________ circuit.

b To measure the current in the cell a ________________ needs to be inserted at point ______.

c The current through the left globe is ______________ compared to the current in the right globe.

d Breaking the circuit at point/s _______ would stop both globes working.

e Breaking the circuit at point/s _______ would stop the right globe from working.

f The positive terminal of the cell is on the ________ side.

4 Find the current in this circuit using Ohm's law.

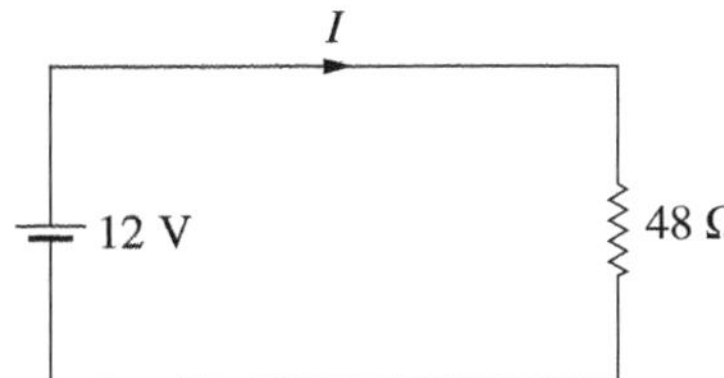

____

____

____

5 Consider the following circuit.

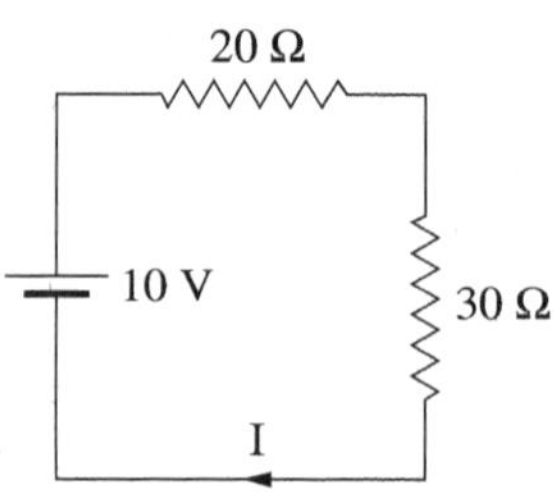

a Is this a series circuit or a parallel circuit?

b Find the size of the current.

6 Find the current and the voltage of the cell in the following circuit.

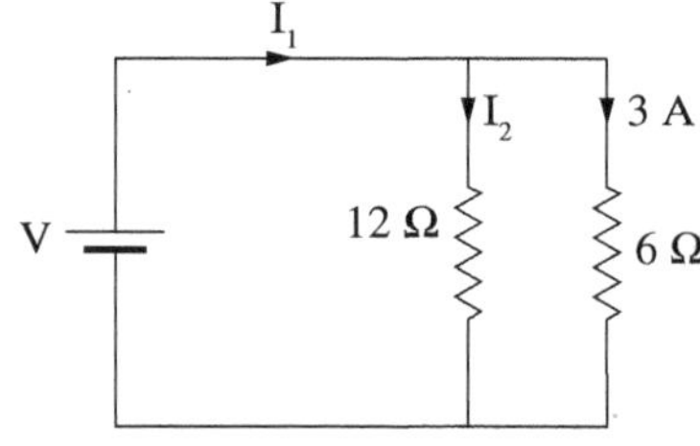

7 The resistance of a conductor depends on three things. What are they?

8 Indicate whether each of the following statements is true (T) or false (F).

a ___ The charge on an electron is equal to the negative charge on a proton.
b ___ An electron in an electric field experiences a force in the direction of the electric field.
c ___ In the right-hand grip rule, the thumb points in the direction of conventional current.
d ___ In conventional current, current is said to flow from the positive terminal to the negative terminal.
e ___ The total resistance in a series circuit equals the sum of the individual resistances in the circuit.
f ___ The magnetic field inside a solenoid depends on the length and diameter of the solenoid.
g ___ An electromagnet creates a single magnetic pole, called a monopole.
h ___ The SI unit for magnetic field strength is the tesla.

 ISBN 978 1 4886 1935 9

WORKSHEET 4.2

Charging it up—static electricity

This worksheet can be done either experimentally or by researching suitable sources.

It is suitable for a home activity because the materials required are likely to be available at home. Check your bathroom, kitchen and bedroom, ask your parents, and consult the manufacturer's tags to confirm the materials.

Insulating materials can be given a static charge by rubbing them with different materials.

Obtain a rod or tube of each of the following materials, which are all insulators:

- glass
- PVC
- acrylic (e.g. Perspex)

Rub a rod of each insulator with the following materials:

- silk
- fur
- wool

1 Determine whether the charge that was produced on each insulator was positive or negative, and complete the table below with your experimental or researched results.

Insulator	Charge with silk	Charge with fur	Charge with wool	
glass				
PVC				
acrylic				

2 An additional row and column have been left in the table. Find another insulator and material, test the static charge produced, and add the results to the table.

3 Without first knowing one of the charges, how can you determine what charge each material/insulator combination produces?

4 Why does a particular combination of insulating materials produce a particular charge (positive or negative)? Can the charge be predicted?

RATING MY LEARNING	My understanding improved	Not confident ◄——► Very confident ○ ○ ○ ○ ○	I answered questions without help	Not confident ◄——► Very confident ○ ○ ○ ○ ○	I corrected my errors without help	Not confident ◄——► Very confident ○ ○ ○ ○ ○

WORKSHEET 4.3

Electric field lines

Using the electric field lines representation, qualitatively model the direction and strength of electric fields produced by:

a a single positive point charge

⊕

b a single negative point charge

⊖

c two positive point charges

⊕ ⊕

d two negative point charges

⊖ ⊖

e a positive and a negative charge (dipole)

⊕ ⊖

f parallel charged plates

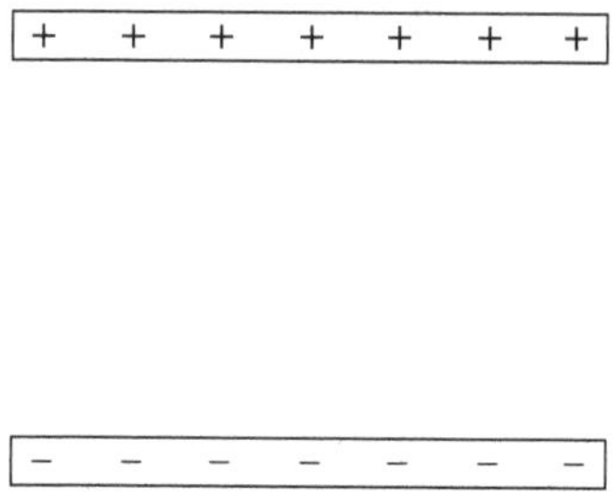

+	+	+	+	+	+	+

–	–	–	–	–	–	–

RATING MY LEARNING	My understanding improved	Not confident ◄──► Very confident ○ ○ ○ ○ ○	I answered questions without help	Not confident ◄──► Very confident ○ ○ ○ ○ ○	I corrected my errors without help	Not confident ◄──► Very confident ○ ○ ○ ○ ○

ISBN 978 1 4886 1935 9

WORKSHEET 4.4

Forces in electric fields

Coulomb's law is used to calculate the size of the electrostatic force acting between charges. When it is applied to two charges, the direction of the force vector is along a line joining the two charges, either a mutual repulsion or attraction. When three or more charges interact, the net force on any one of them is the vector sum of the forces applied by the other charges.

The electric field is defined as the region in space in which an electric force would be experienced by a charge, given by the equation $\vec{E} = \frac{\vec{F}}{q}$.

Electric fields can be represented qualitatively by means of field lines, where the direction of the field is a tangent to the field line itself and the field strength can be judged by the 'density' of the field lines.

A more precise representation of the field at a point is obtained by the application of the equation above. If the charge is 1 coulomb, then the field can be thought of as the force on a unit charge.

1 In the diagram below there is a charge $Q = +5\text{ C}$ at (0,0) and a 'test' charge q_1 of +1 C placed 3 m away at (0,3).

The force on the test charge is represented by the vector $\vec{F}$.

In the diagram, take the positive y-direction as north and the positive x-direction as east.

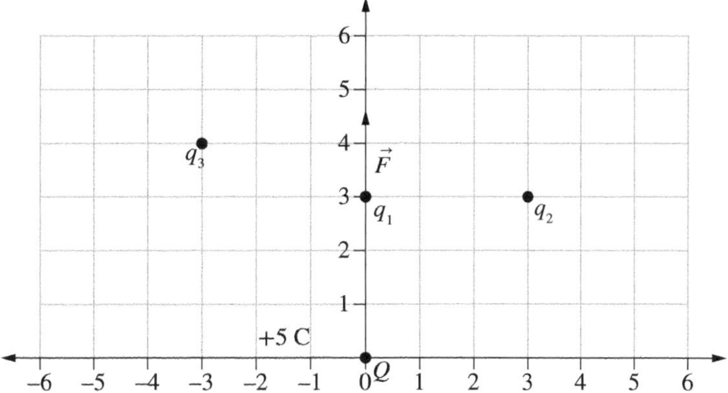

a Calculate the magnitude of $\vec{F}$.

b What is the magnitude of the electric field E at q_1? Include the correct units.

c q_2 is also a +1 C test charge. It has been placed at (3,3). Calculate the size of the electric field at that point due to its interaction with charge Q, and fully express the electric field vector.

Draw the corresponding electric field vector. The size of the vector you draw is not important, but the direction is very important.

d Repeat part c for q_3, another test charge of +1 C at (–3,4).

2 Consider the situation when two charges are creating the field being tested.

In the diagram below two charges Q_1 and Q_2, each of +5 C, are placed at coordinates (–3,0) and (3,0) respectively.

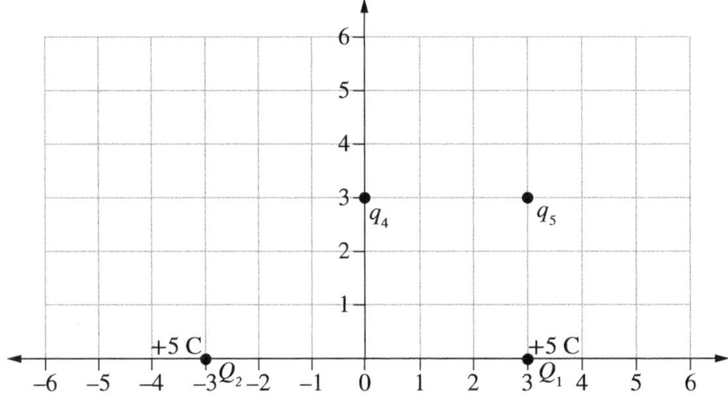

The electric field acting on charge q_4 = +1 C at (0,3) is the vector sum of the fields produced by Q_1 and Q_2. Call these fields $\vec{E}_{Q1}$ and $\vec{E}_{Q2}$.

a Find the size and direction of $\vec{E}_{Q1}$ and $\vec{E}_{Q2}$, and draw them on the diagram. Then find the net electric field, $\vec{E}_4$.

b Repeat part **a** for the charge q_5 = +1 C at (3,3).

c The charge at (3,0) is now replaced by a charge of –5 C. Describe what happens to the field at (3,3).

RATING MY LEARNING	My understanding improved	Not confident  Very confident ○ ○ ○ ○ ○	I answered questions without help	Not confident  Very confident ○ ○ ○ ○ ○	I corrected my errors without help	Not confident ↔ Very confident ○ ○ ○ ○ ○

 ISBN 978 1 4886 1935 9

WORKSHEET 4.5

Series and parallel

Find the unknown quantities in each of the following circuits. Show your working in each case.

P = power, I = current, R = resistance and V = EMF

1

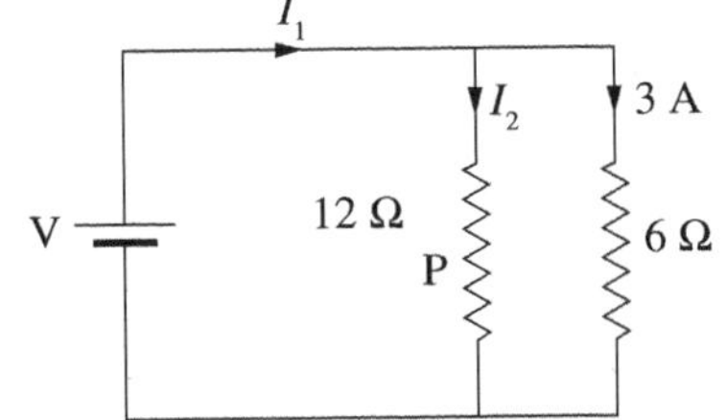

2

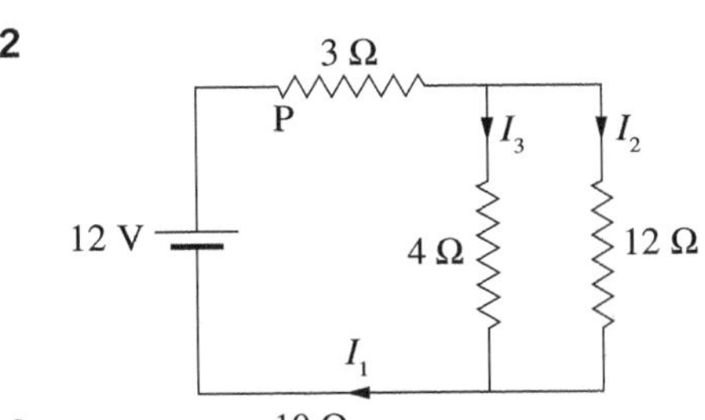

3

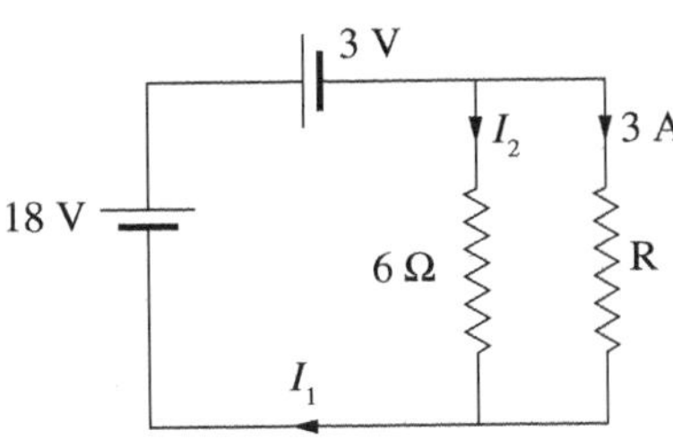

4

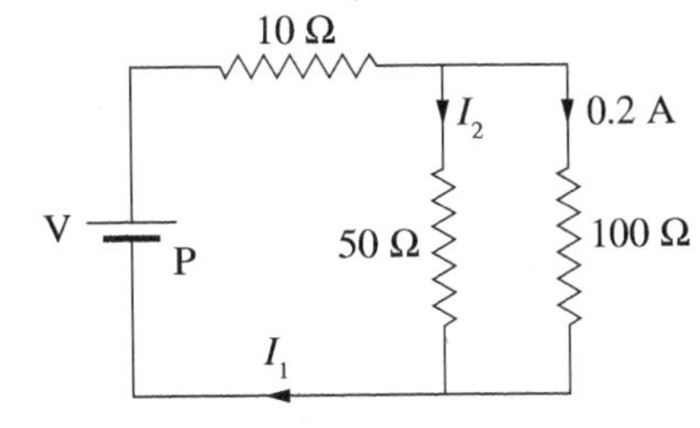

1

2

3

4

RATING MY LEARNING	My understanding improved	Not confident ◄──► Very confident ○ ○ ○ ○ ○	I answered questions without help	Not confident ◄──► Very confident ○ ○ ○ ○ ○	I corrected my errors without help	Not confident ◄──► Very confident ○ ○ ○ ○ ○

ISBN 978 1 4886 1935 9

WORKSHEET 4.6

Resistance is variable—Ohm's law

Consider a simple circuit that was constructed to test whether particular circuit loads are ohmic or non-ohmic. Ohmic circuits obey Ohm's law (i.e. $V = IR$).

The circuit consisted of some patch cords, a variable power supply, a resistor, and voltage and current sensors.

The voltage was varied through three different resistors in turn, and the following voltage–current graph was drawn.

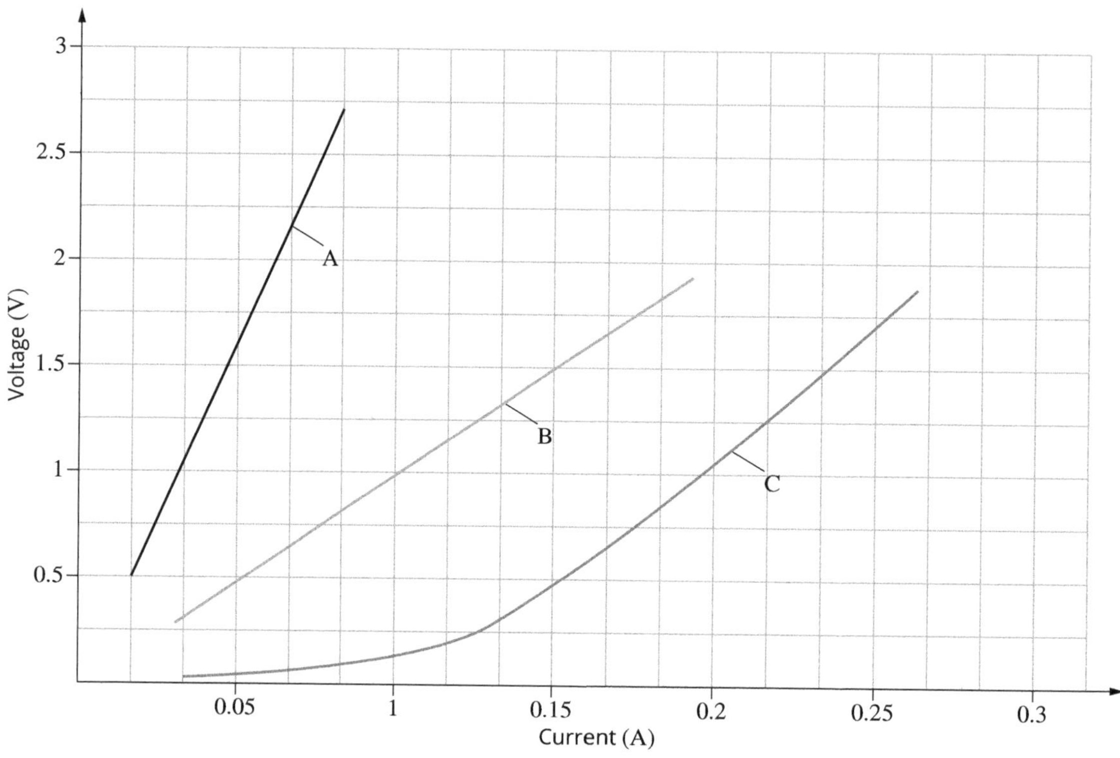

1 Determine whether each resistor is ohmic or non-ohmic, and circle your answer.

Resistor A	ohmic	non-ohmic
Resistor B	ohmic	non-ohmic
Resistor C	ohmic	non-ohmic

2 Explain the reasoning behind your answer to Question 1.

3 Using the V–I graph above, determine the resistance of resistors A and B.

4 Calculate the resistance of the non-ohmic resistor as the voltage is varied from 0.50 V to 1.9 V.

ISBN 978 1 4886 1935 9

5 For what conditions is your answer to Question 4 valid for resistor C? Does it apply to any voltage?

6 The voltage across a 580 Ω ohmic resistor is 120 V. What is the current?

7 The current through a 100 Ω ohmic resistor is 0.150 A. What voltage is being applied?

8 A circuit with a 3.0 V battery pack and an ohmic resistor has a current of 0.060 A. What is the value of the resistor?

RATING MY LEARNING	My understanding improved	Not confident ◄—► Very confident ○ ○ ○ ○ ○	I answered questions without help	Not confident ◄—► Very confident ○ ○ ○ ○ ○	I corrected my errors without help	Not confident ◄—► Very confident ○ ○ ○ ○ ○

WORKSHEET 4.7

Heating and power loss in electric circuits

1 Heat loss in electric circuits can be a positive benefit in some circumstances and negative in others. From your studies of energy and power loss in electric circuits, list some examples where heating has a positive effect and others where it has a negative effect.

Positive effect	Negative effect

2 100 J of heat is produced each second in a 10 Ω resistor. What is the potential difference across the resistor?

3 An electric iron has a resistance of 24 Ω and uses a current of 10 A. Calculate the total heat the iron transfers to its surrounds each minute.

4 An electric stove top uses energy at a rate of 1500 W at its maximum setting. At the minimum setting it uses energy at a rate of 350 W. If the voltage supplied is 240 V, what is the current and the resistance in each case?

5 A large electric room heater with a resistance of 16 Ω draws 15 A when heating at its maximum setting. If it runs for 2 hours, how much energy has been used? Give your answer in both kilowatt hours and joules.

ISBN 978 1 4886 1935 9

6 Two resistance wires, each of 10 Ω, are connected across a battery of 10 V. Calculate:

a The power supplied through the wires when connected in series.

b The power supplied through the wires when connected in parallel.

7 What is the ratio of the power supplied via a series circuit to that supplied by a parallel circuit?

8 Two options are being considered in designing a new hair dryer. Both use the same length of the same resistive wire. In the first option the wire is connected in series, while the second option the wires are connected in parallel, which gives a more compact dryer. The designers want to produce the highest power output possible. Calculate the ratio of the power output of the series option compared with the parallel option, to advise the designers about which option they should choose.

RATING MY LEARNING	My understanding improved	Not confident ◄—► Very confident ○ ○ ○ ○ ○	I answered questions without help	Not confident ◄—► Very confident ○ ○ ○ ○ ○	I corrected my errors without help	Not confident ◄—► Very confident ○ ○ ○ ○ ○

ISBN 978 1 4886 1935 9

WORKSHEET 4.8

Magnetic fields

1 Three bar magnets are arranged end to end, as shown in the diagram below. Indicate, using an arrow pointing inwards or outwards, where the magnets will be repelling and where they will be attracting.

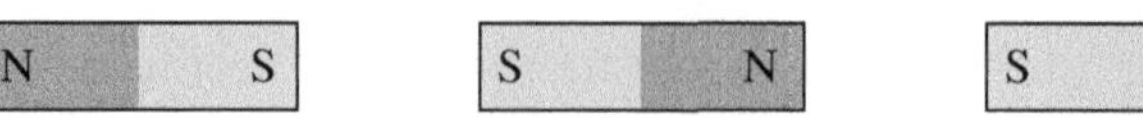

Draw the magnetic field between each of the following combinations of bar magnets.

2

3

4

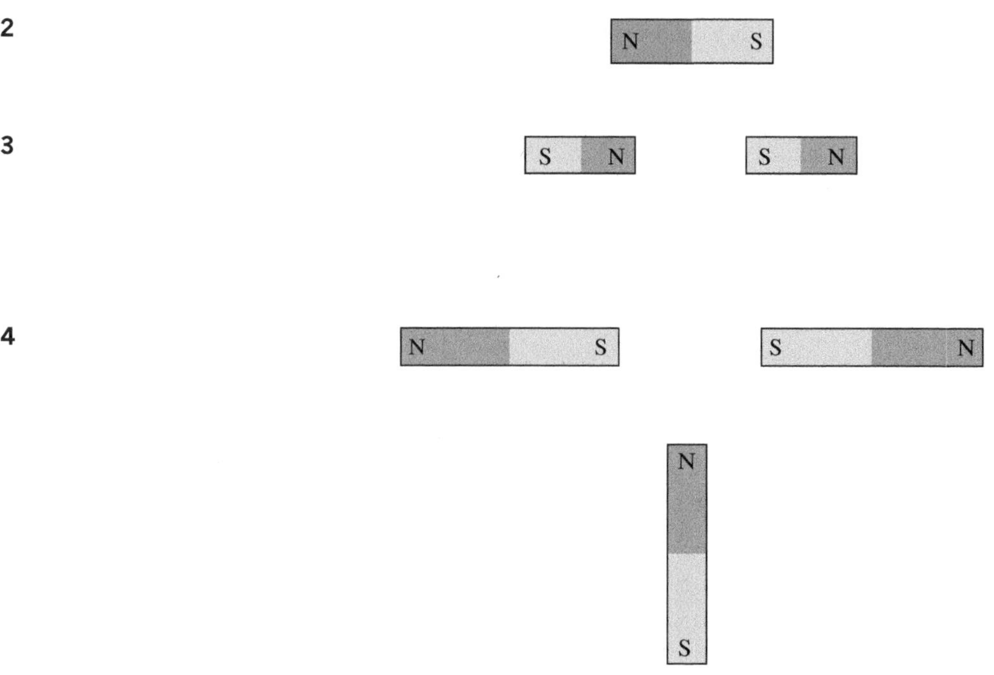

The circles in the following diagrams represent compasses. For each arrangement, draw in the needle of the compass, indicating the direction that it would be pointing.

5

6

7

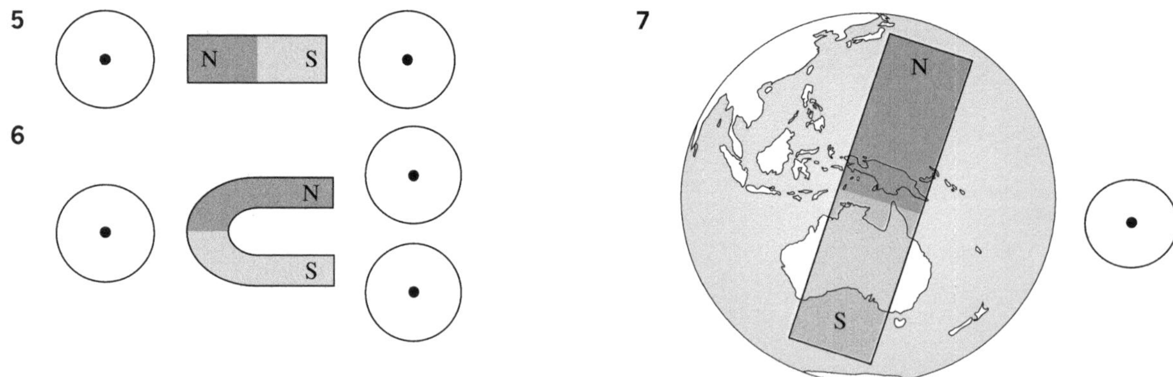

8 What rare event must have happened to make the figure in Question 7 correct?

RATING MY LEARNING	My understanding improved	Not confident ◄——► Very confident ○ ○ ○ ○ ○	I answered questions without help	Not confident ◄——► Very confident ○ ○ ○ ○ ○	I corrected my errors without help	Not confident ◄——► Very confident ○ ○ ○ ○ ○

 ISBN 978 1 4886 1935 9

WORKSHEET 4.9

Making it magnetic—fields and coils

1 What makes a material magnetic? Are there any naturally occurring minerals that are magnetic?

2 The needle in a compass acts like a tiny bar magnet. When the needle is in a strong magnetic field, it will align itself with that field. If the black end of the needle represents the south pole of a magnet, which direction would the black end point if the compass were placed between the poles of a horseshoe magnet?

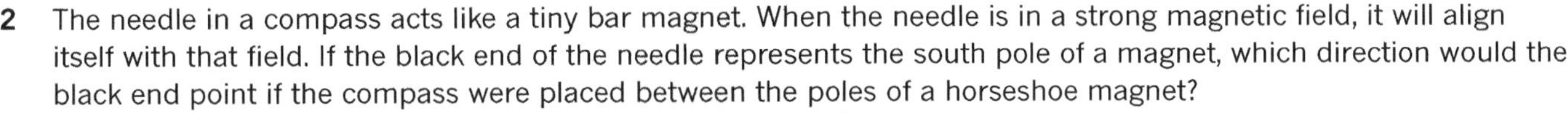

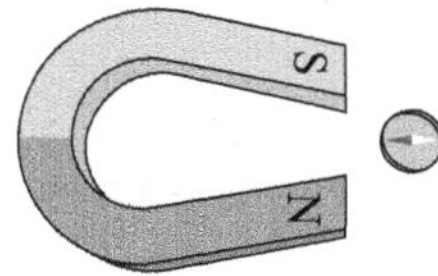

3 If you had two identical bar magnets separated by distance d with the north pole of each magnet facing the other, what direction would a compass point at distance of $0.5\,d$ (i.e. exactly half-way between the magnets)? Explain.

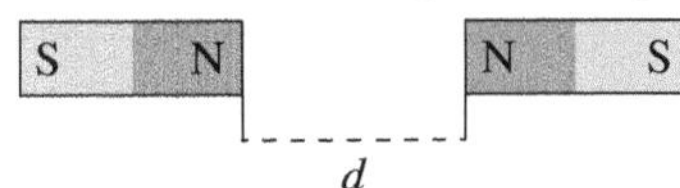

4 Two magnets each produce a magnetic field strength of 0.25 T at a distance $0.5\,d$. If the north and south poles of the magnets are facing each other a distance d apart, what is the magnetic field strength half-way between these two poles?

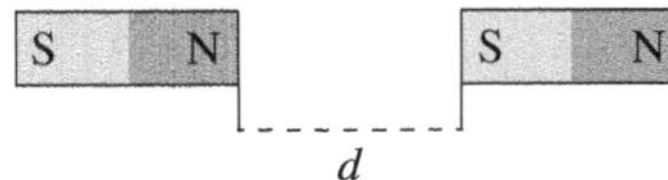

5 In the loop of wire shown in the figure below, the current flows anticlockwise. In what direction is the magnetic field inside the loop?

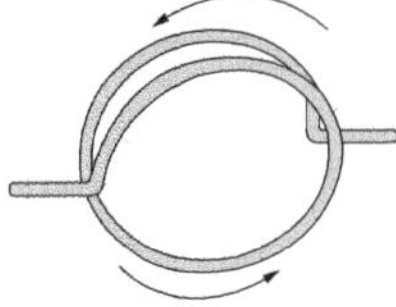

6 In the figure below, the coil on the right has a radius half that of the coil on the left. What must happen to the current in the loop on the right in order to induce the same magnetic field strength in the centre of the coil, compared the loop on the left?

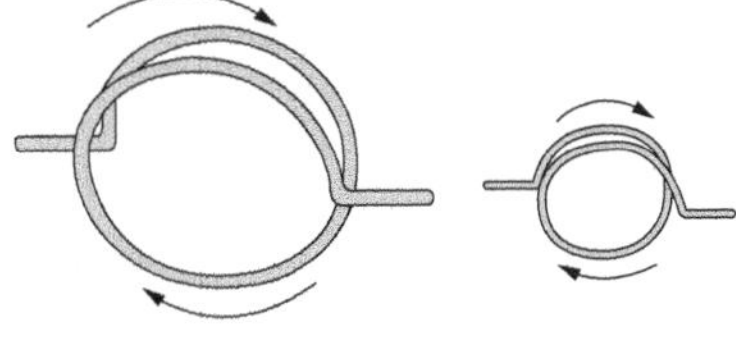

RATING MY LEARNING	My understanding improved	Not confident ◄——► Very confident ○ ○ ○ ○ ○	I answered questions without help	Not confident ◄——► Very confident ○ ○ ○ ○ ○	I corrected my errors without help	Not confident ◄——► Very confident ○ ○ ○ ○ ○

ISBN 978 1 4886 1935 9

WORKSHEET 4.10

Literacy review—building your vocabulary

1 An important convention is followed for the naming of units and quantities after people, compared with other terms. Looking back over the modules covered this year, what is this convention?

____

2 The following is a list of units and quantities relating to electricity and magnetism. A significant proportion of these units are named after people. Categorise the quantities and units shown as being named after people or derived from a related term.

volt current ampere resistance ohm solenoid

magnetism tesla gauss power energy watt joule

Honouring a past scientist	Derived from a related term

3 For the units and quantities you listed in Question 2 as being derived from a related term, write the meaning and the origin of the word in the table below.

Word	Meaning	Origin

4 The following statement talks about series and parallel circuits. Fill in the blanks from the list of randomly ordered words. Some words may be used more than once.

current resistance Ohm's law inversely directly

For a circuit with constant ______________, as voltage increases or decreases, ______________ increases or decreases respectively. They are said to be ______________ proportional. For the same circuit, when resistance increases, ______________ decreases, and vice versa. They are said to be ______________ proportional. This relationship is called ______________.

5 Fill in the blanks from the following random list of terms related to magnetic fields in coils. Some words may be used more than once.

permanent solenoid radius inversely current

directly turns magnetic field strength electromagnets

Two main types of magnets are ______________ magnets and ______________. A wire carrying a ______________ induces a magnetic field around it. When the wire is wound into a circle, it is called a coil or ______________. The ______________ at the centre of a coil depends upon the ______________ and the ______________ of the coil. If the coil contains many ______________, the magnetic field strength will ______________. The current is ______________ proportional to the ______________. The ______________ is inversely proportional to the ______________.

RATING MY LEARNING	My understanding improved	Not confident ◄——► Very confident ○ ○ ○ ○ ○	I answered questions without help	Not confident ◄——► Very confident ○ ○ ○ ○ ○	I corrected my errors without help	Not confident ◄——► Very confident ○ ○ ○ ○ ○

 ISBN 978 1 4886 1935 9

WORKSHEET 4.11

Thinking about my learning

On completion of Module 4: Electricity and magnetism, you should be able to describe, explain and apply the relevant scientific ideas. You should be able to work with data, to interpret, analyse and evaluate it.

1 The table lists the key knowledge covered in this module. Read each and reflect on how well you understand each concept. Rate your learning by shading the circle that corresponds to your level of understanding for each concept. It may be helpful to use colour as a visual representation. For example:

- green—very confident
- orange—in the middle
- red—starting to develop.

Concept focus	Rate my learning				
	Starting to develop ◄				► Very confident
How objects are electrically charged	○	○	○	○	○
Electric field lines	○	○	○	○	○
Electrostatic force and Coulomb's law	○	○	○	○	○
Electric potential energy and its relationship to the electric field	○	○	○	○	○
Current and the difference between electron flow and conventional current	○	○	○	○	○
Ohm's law	○	○	○	○	○
Analysis of series circuits	○	○	○	○	○
Analysis of parallel circuits	○	○	○	○	○
Ferromagnetic materials	○	○	○	○	○
Magnetic field lines	○	○	○	○	○
The relationship between electric current and magnetic fields. This includes both current-carrying wires and solenoids.	○	○	○	○	○

2 Consider points you have shaded from starting to develop to middle-level understanding. List specific ideas you can identify that were challenging.

3 Write down two different strategies that you will apply to help further your understanding of these ideas.

PRACTICAL ACTIVITY 4.1

Charge and electric field

Suggested duration: 30+ minutes

INTRODUCTION

Individual atoms normally have a neutral charge, meaning that the number of electrons equals the number of protons. If the number of protons exceeds the number of electrons, the atom has a positive charge. Conversely, if the number of electrons exceeds the number of protons, the atom has a negative charge.

Objects with like charges repel one another, and objects with unlike charges attract one another.

In the right conditions, an object can acquire an electric charge by being rubbed against another object. An example is rubbing a balloon against your hair. The charges on the balloon and hair will be the same magnitude but opposite in polarity, so the balloon and hair attract each another. Rubber and plastic are insulators, which means that charge does not move easily through them.

In this experiment, a Faraday ice pail is used to examine the transfer of electric charges from one material to another.

MATERIALS

- Van de Graaff generator
- plain balloons
- metal-coated balloons (available from novelty shops and supermarkets)
- woollen cloth
- string
- foil pie plates
- discharging sphere
- insulating rod
- galvanometer (or charge sensor and data acquisition system)
- rubber mat to stand on

PURPOSE

To investigate and describe processes by which materials become electrically charged, and the factors that affect how charged objects interact at a distance.

 A Van de Graaff generator creates a very high voltage, but the amount of actual charge is relatively low, making it completely safe when used responsibly.

People with heart conditions should not participate in activities involving Van de Graaff generators.

Sensitive electronic equipment such as phones, tablets and computers should be kept well away from the generator.

PROCEDURE

Balloon test 1

1 Charge a standard rubber balloon suspended by a string by rubbing it with a woollen cloth (or woollen jumper).

2 Start the generator and bring a coated balloon near the sphere, held by an insulating rod. Carry the coated balloon to the suspended charged balloon. Observe and describe the action between them.

3 Record your observations in the Data and analysis table below.

Balloon test 2

1 Fix two coated balloons to the side of the sphere of the generator with damp cotton strings. Start the generator. Observe and describe the action between the balloons.

2 Record your observations in the Data and analysis table below.

Galvanometer or charge sensor

1 Following the instructions that come with the generator, connect a galvanometer or charge sensor capable of registering small currents in series with the generator and an earth. If you have access to a data acquisition system, set the sample rate to between 200 Hz and 500 Hz.

 ISBN 978 1 4886 1935 9

2 Hold the discharge sphere 1–2 cm from the generator's sphere. Turn the galvanometer to its highest range and start the generator.

3 Observe the discharge and the reading on the galvanometer. If no reading is observed, turn to a slightly more sensitive range and repeat.

4 Record your observations in the Data and analysis section below.

Pie plates

1 Make sure the sphere of the generator is fully discharged.

2 Stack a group of foil pie plates upside down, and loosely, on top of the sphere. Start the generator. Observe and describe what happens to the pie plates. Record your observations in the Data and analysis table below.

Other tests

Check with your teacher as to whether other tests are possible with accessories the school has for the Van de Graaff generator. Record your observations of one of these tests in the extra space in the Data and analysis table below.

DATA AND ANALYSIS

Test	Observations
Balloon test 1	
Balloon test 2	
Galvanometer (or charge sensor)	
Pie plates	

ISBN 978 1 4886 1935 9

CONCLUSION

1 Why should the cotton strings in Balloon test 2 be damp? Why couldn't they be nylon fishing line?

2 Examine the construction of the Van de Graaff generator. Explain the method by which charge is transferred to the sphere. (The top of the sphere can usually be carefully removed to view the internal operation. If not, find a diagram on the internet.)

3 Describe the distribution of charge on the sphere.

4 Use your observations in this practical activity to describe the processes by which objects become electrically charged.

5 The brush and belt of the generator charge the belt by friction. The sphere at the top collects the charge. What function does the moving belt have? Why must it be an insulated belt?

6 Is there any basic difference between the charge from friction (rubbing a woollen cloth or a balloon, for example) and that by a Van de Graaff generator?

7 Can charges obtained from friction (for example electrostatic charge from a Van de Graaff generator) form a current?

RATING MY LEARNING	My understanding improved	Not confident ←→ Very confident ○ ○ ○ ○ ○	I answered questions without help	Not confident ←→ Very confident ○ ○ ○ ○ ○	I corrected my errors without help	Not confident ←→ Very confident ○ ○ ○ ○ ○

 ISBN 978 1 4886 1935 9

PRACTICAL ACTIVITY 4.2

Mapping electrical fields

Suggested duration: 55 minutes

INTRODUCTION

All charged objects produce electric fields in the space surrounding them. If you know the shape, direction and magnitude of an electric field, you can determine how a charged particle will interact with the field.

Electric field lines are a convenient visual way of representing an electric field. They are lines drawn to follow the path of the electric field, originating from a positive charge (or charged object) and ending at a negative charge (or charged object). The lines never cross, and the density of lines (the closeness of the lines) represents the magnitude of the field strength. Lines of equal electric potential are known as isolines.

In this activity you will identify different isolines of electric potential surrounding a pair of charged electrodes, then use those isolines as guides to draw electric field lines and identify the magnitude and direction of the electric field.

MATERIALS

- semi-conductive paper with grid
- conductive ink pen
- metal drawing pins
- metal T-pin
- cork board
- multimeter or voltage sensor
- banana plug patch cords with alligator clips
- pencil
- white or light-coloured felt tip marker
- power supply, 18 V, 3 A

PURPOSE

Using the principles of electric fields and electric potential energy, investigate the lines of equal electric potential surrounding oppositely-charged electrodes and the shape and direction of the corresponding electric field lines.

 Do not connect the terminals of a power supply without a load or resistance. This will cause a short circuit, which may damage the power supply.

PROCEDURE

Drawing dipole electrodes with conductive ink

1 Place a sheet of semi-conductive paper flat, printed side up, on the lab bench.

2 Use the pencil to draw two circles 10 cm apart along the horizontal centre line of the paper. Make each dot about 1 cm in diameter.

3 Shake the conductive ink pen (with the cap on) for 10 to 20 seconds. Then remove the cap and press the spring-loaded tip down on the semi-conductive paper, in the centre of either circle.

4 Lightly squeeze the barrel on the pen until ink starts to slowly flow onto the paper, then move the tip of the pen around inside the circle until the circle is completely filled with the ink.

5 If the dot is not completely filled in, draw over it again with the pen. A solid, uniform dot is essential for good measurements.

6 Repeat the same process for the second circle.

Allow plenty of time for the conductive ink to dry completely. During this time, move on to the next step.

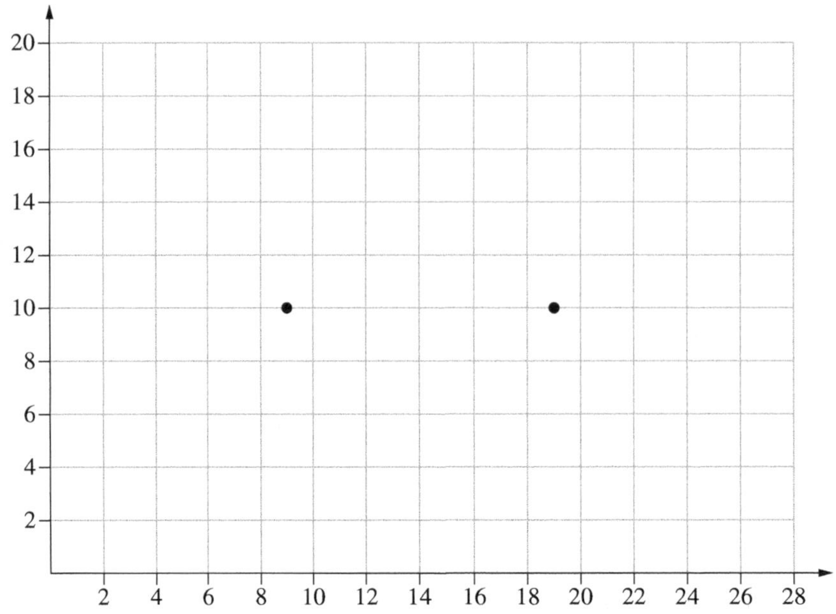

Drawing parallel-plate electrodes with conductive ink

1 Place a sheet of semi-conductive paper flat, printed side up, on the lab bench.

2 Use the ruler and pencil to draw two straight lines 10 cm long and 8 cm apart, and perpendicular to the horizontal centre line of the paper. Make each line about 5 mm wide.

3 Shake the conductive ink pen (with the cap on) for 10 to 20 seconds. Then remove the cap and press the spring-loaded tip down on the semi-conductive paper, in the centre of either line.

4 Lightly squeeze the barrel on the pen until ink starts to slowly flow onto the paper, and then move the tip of the pen back and forth inside the sketched line until it is completely filled with the ink.

5 Use care when drawing the electrodes. If the line is not uniformly thick, it may not conduct. Go back and deposit more conductive material if necessary. If the line is not completely filled in, draw over it again. A solid uniform line is essential for good measurements.

6 Repeat the same process for the second line.

Allow plenty of time for the conductive ink to dry completely. This may take overnight.

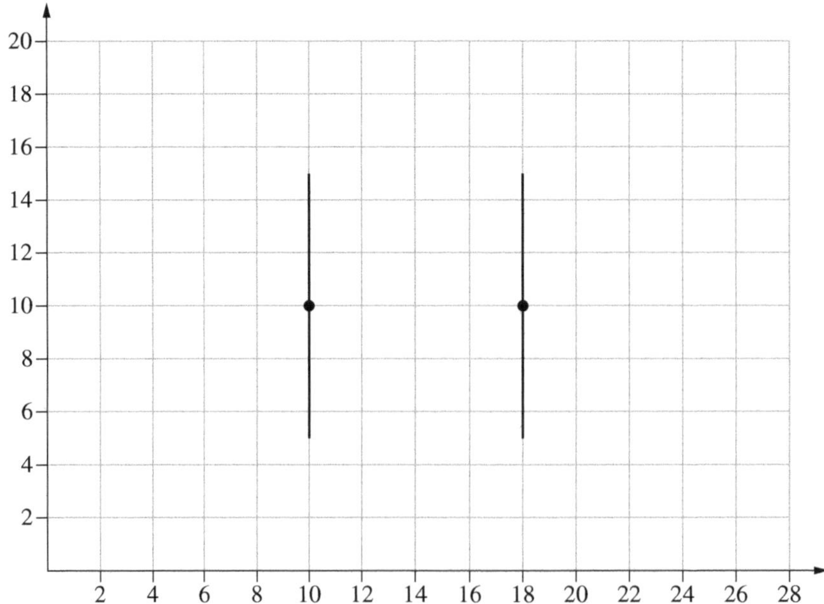

 ISBN 978 1 4886 1935 9

PRACTICAL ACTIVITY 4.2

Dipole electric fields

1 Place the first sheet of semi-conductive paper with the dipole electrodes drawn on it onto the cork board. Pin each corner of the paper to the cork board using metal drawing pins.

2 Press a drawing pin into the centre of each electrode, making sure that the pins are pressed firmly through the paper and into the corkboard.

3 With the power supply off, connect the electrode pins to the terminals on the power supply using two patch cords and alligator clips, as shown in the diagram below. The positive terminal is connected to one electrode (which will be the positive electrode), and the negative terminal is connected to the other electrode (which will be the negative electrode).

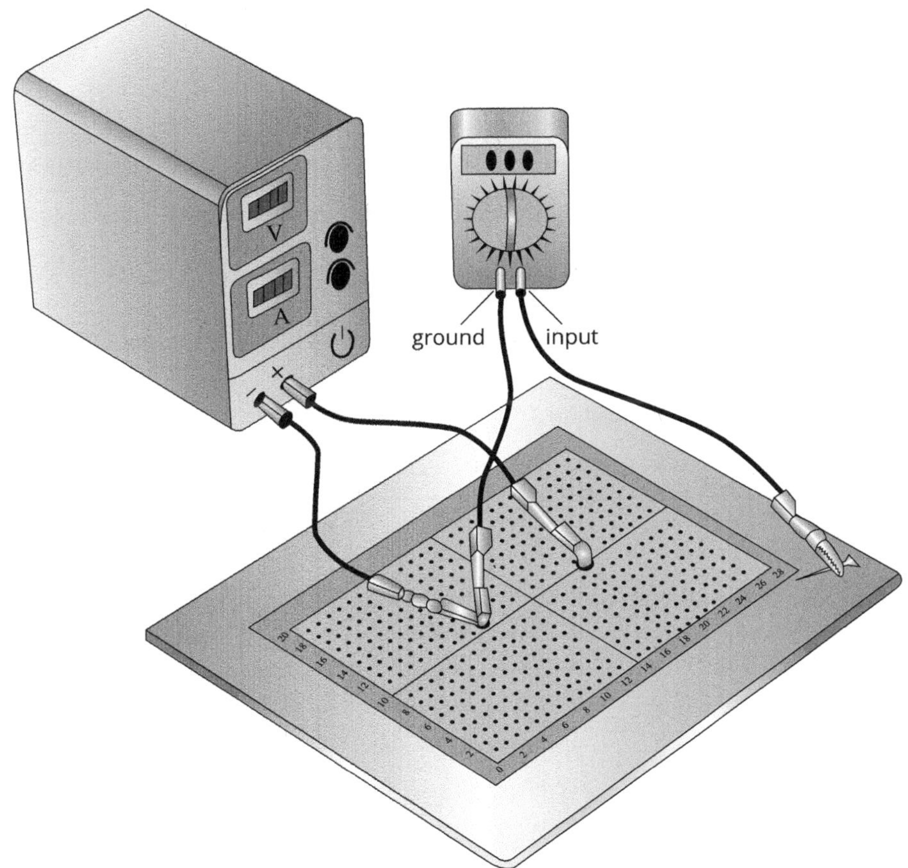

4 Connect the two remaining patch cords to the DC voltage ports on the digital multimeter or voltage sensor, and then adjust the meter to measure DC voltage up to 10 V DC.

5 Attach alligator clips to the ends of the meter patch cords. Clip the meter 'ground' or 'COM' patch cord to the pin in the negative electrode; this is now the reference electrode. Connect the alligator clip on the other patch cord to the metal T-pin.

6 Turn on the power supply and adjust the voltage to 10 V DC.

7 Touch the tip of the T-pin to several solid black areas on the paper, and observe the voltage measurement on the meter. Repeat for a few other solid black areas. The meter should show different voltages at different points on the paper. If not, make sure all of the alligator clips and drawing pins are making good connections, then retest.

> Touch the tip of the T-pin only to the solid black areas of the semi-conductive paper. Do not touch it to the paper's grid marks.

8 Use the tip of the T-pin to test different positions surrounding the electrodes until you find a location on the paper where the meter reads 1.0 V. Use the felt-tip marker to make a small mark at this position.

ISBN 978 1 4886 1935 9

9 Continue moving the T-pin to different points on the paper, identifying several positions surrounding the electrodes where the voltage is at 1.0 V. Mark each new point with the felt-tip marker until there are enough marks to accurately draw a smooth line that connects them. Label the line '1.0 V'.

 Be sure to probe all areas surrounding the electrodes. Isolines of electric potential may connect in a closed path on the paper or extend off the edge of the paper and then re-enter the paper in an unexpected location.

10 Repeat the previous data collection steps, identifying and drawing the isolines of electric potential for 3.0 V, 5.0 V, 7.0 V and 9.0 V. Label each line with its corresponding voltage.

11 Turn the power supply off when you are finished.

Parallel plates

Using the second sheet of semi-conductive paper with the two parallel-plate electrodes drawn on it, repeat the setup procedure outlined in Part 1. For each line (plate), push the drawing pin into the plate about 10 mm from one end.

1 Clip the DMM ground patch cord to the drawing pin in the negative plate electrode; this is now the reference electrode.

2 Turn on the power supply and adjust the voltage to 10 V DC.

3 Following the Part 1 data collection steps, identify and draw the isolines of electric potential surrounding the parallel-plate electrodes for 1.0 V, 3.0 V, 5.0 V, 7.0 V and 9.0 V.

4 Turn the power supply off when you are finished.

DATA AND ANALYSIS

Dipoles

1 Using the pencil, draw a line starting from any point on the surface of the positive electrode and extending to the nearest isoline. Draw the line so it leaves the surface of the electrode at a right angle and intersects the first isoline at a right angle. The field line must curve smoothly as shown below.

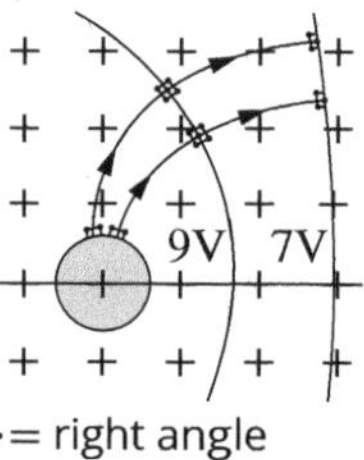

·= right angle

2 When you are satisfied with the shape of the line, draw over it using the felt-tip pen, and then draw an arrow head on the line indicating the direction of the electric field.

3 From the tip of the arrow you have just drawn, draw another field line like the previous one extending from that isoline line to the next, and then another field line to the next isoline, and so on until you reach the other electrode or the edge of the paper. Combine the connecting lines to form one electric field line.

4 Repeat these steps to draw a total of 10 electric field lines originating from the positive electrode, spacing the start of each line evenly along its surface.

Parallel plates

Repeat the procedure from the dipoles to draw 10 electric field lines from the positive parallel-plate electrode to the negative parallel-plate electrode, as follows:

1 Draw 7 of the 10 field lines leaving the inner surface of the positive electrode, and the remaining 3 field lines leaving the outer surface of the electrode.

2 Space the 7 lines evenly along the inner surface, and the 3 lines evenly along the outer surface of the positive electrode. Draw each line with a smooth curve when necessary, and include arrow heads indicating the direction of the electric field.

 ISBN 978 1 4886 1935 9

PRACTICAL ACTIVITY 4.2

CONCLUSION

1 What is the convention for showing the direction of electric fields?

2 Based on your final electric field plots, what are the primary similarities and differences between the electric fields surrounding each electrode configuration?

3 Where was the electric field strongest in each electrode configuration? Justify your answer based on your results.

4 What can be said about the voltage at each point of an isoline of electric potential?

5 What do the electric field lines represent?

6 How would the shape of the electric field lines in each configuration change if you increased the voltage across the electrodes? Justify your answer.

RATING MY LEARNING	My understanding improved	Not confident ◄——► Very confident ○ ○ ○ ○ ○	I answered questions without help	Not confident ◄——► Very confident ○ ○ ○ ○ ○	I corrected my errors without help	Not confident ◄——► Very confident ○ ○ ○ ○ ○

PRACTICAL ACTIVITY 4.3

Ohmic and non-ohmic conductors

Suggested duration: 30 minutes

INTRODUCTION

An ohmic conductor has a resistance that obeys Ohm's law. That is, its resistance does not vary but is a constant, so that $V \propto I$. Non-ohmic conductors have a resistance that is not constant: it changes as the current or potential difference changes.

PURPOSE

To investigate the properties of an ohmic and a non-ohmic conductor.

 Most current sensors have a 1 A range and are suitable for this experiment. They are generally current-protected, but check the manufacturer's specifications and use the equipment with care. Exceeding the maximum current rating will permanently damage an ammeter or sensor.

Exercise caution when using the power supply: use only low voltages (10 V DC or less) and only make changes to the circuit when the circuit switch is open.

To reduce the risk of electrical shock, do not allow food or drinks near the equipment.

Be sure resistor ratings and power supply settings are appropriate for your voltage and current sensors or meters.

MATERIALS

- power supply 12 V DC, 1 A
- connecting wires
- DC ammeter (0–1 A and 0–5 A) or current sensor
- DC voltmeter (0–15 V) or voltage sensor
- rheostat or variable resistor
- 50 cm length of nichrome wire of known diameter/size
- 12 V light globe and holder

PROCEDURE

1 Connect the non-variable terminals of the rheostat to the DC terminals of the power supply. Connect the remainder of the circuit as shown in the diagram. The positive terminal of the ammeter should be connected to the positive side of the rheostat. Always connect the ammeter to the highest-range scale first (0–5 A). If the current reading is too small, then connect to the lower scale (0–1 A).

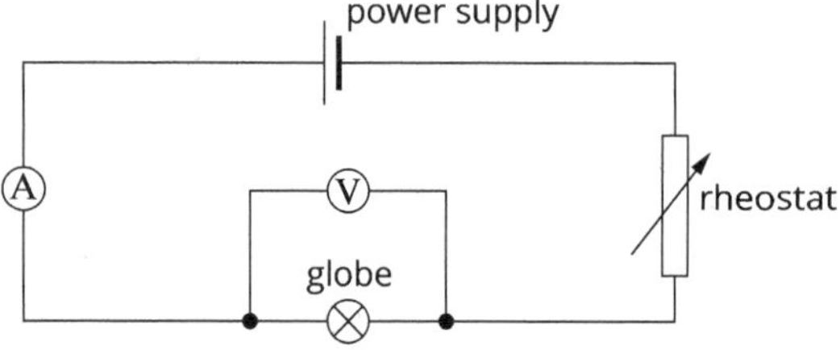

2 Connect the voltmeter initially across the globe. Set the sliding contact of the rheostat near the positive end to produce the rheostat's lowest resistance value. Turn the power supply to 6 V and watch the meters. What change do you observe with the circuit connected correctly? Record your observations in the Data and analysis table.

3 Adjust the rheostat until a current of 0.2 A flows through the globe. Record the corresponding potential difference across the globe in the Data and analysis table.

4 Using the rheostat, increase the current in steps of 0.2 A. Record the potential difference and current each time in the Data and analysis table.

5 Switch off the power supply and replace the globe in the circuit with a 50 cm length of nichrome wire stretched out reasonably straight. You may find it best to tape it to a wooden ruler, but make sure it is not in contact with any other conductor.

6 Turn the power supply back on and repeat steps 3 and 4 for the nichrome wire, recording your results in the Data and analysis table.

ISBN 978 1 4886 1935 9

DATA AND ANALYSIS

1 What change did you observe with the circuit connected correctly?

2 Record your results in the table below.

Current (A)	Voltage (V)
Globe	
0.2	
0.4	
0.6	
0.8	
1.0	
Nichrome wire	
0.2	
0.4	
0.6	
0.8	
1.0	

3 Plot a graph of V versus I for the globe and the nichrome wire on the same set of axes, using the grid below.

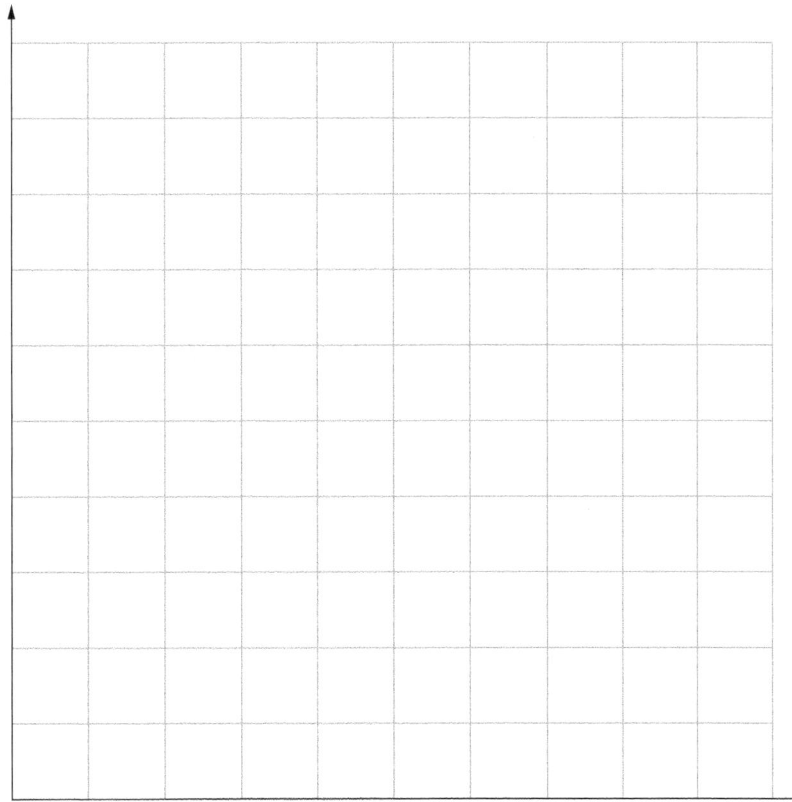

ISBN 978 1 4886 1935 9

CONCLUSION

1 Based upon your results, are the globe and the nichrome wire ohmic or non-ohmic conductors? Make reference to your graph in justifying your conclusion.

2 Determine the value of the resistance for the nichrome wire.

Your teacher may have other conductors for you to test. Repeat the investigation for any additional conductors.

RATING MY LEARNING	My understanding improved	Not confident ◄──► Very confident ○ ○ ○ ○ ○	I answered questions without help	Not confident ◄──► Very confident ○ ○ ○ ○ ○	I corrected my errors without help	Not confident ◄──► Very confident ○ ○ ○ ○ ○

 ISBN 978 1 4886 1935 9

PRACTICAL ACTIVITY 4.4

Series and parallel circuits

Suggested duration: 50 minutes

INTRODUCTION

Ohm's law states that the voltage drop across an electrical resistor is equal to the magnitude of the current through the resistor multiplied by the resistance, i.e. $V = IR$.

When circuits have multiple resistors hooked up in either series or parallel, each resistor will have a voltage and a current associated with it. Can we simplify a complex circuit by substituting one resistor that behaves like several resistors together?

By measuring individual components and comparing them to the behaviour of the whole circuit, it is possible to experimentally determine whether there is an equivalent resistance for a set of resistors.

PURPOSE

To investigate combinations of resistors in series and parallel circuits.

Exercise caution when using the power supply: use only low voltages (10 V DC or less), and only make changes to the circuit when the circuit switch is open.

To reduce the risk of electrical shock, do not allow food or drinks near the equipment.

Be sure resistor ratings and power supply settings are appropriate for your voltage and current sensors or meters.

MATERIALS

- data collection system and current and voltage sensors (or ammeter and voltmeter)
- 3 resistors with different known resistances, e.g. ceramic resistors 10–60 Ω, 5 W
- DC power supply (10 V, 1 A minimum)
- single-pole, single-throw switch
- 4 mm patch cords
- alligator clip adapters for the patch cords

PROCEDURE

If you are using a data collection system, connect the voltage and current sensors and set up a display that enables monitoring of both readings simultaneously.

Resistors in series

1 With the power off, connect the first resistor and the open switch in series to the power supply using the patch cords.

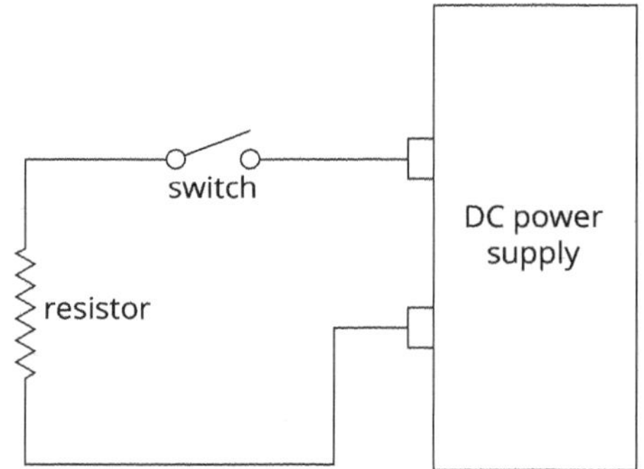

2 Connect the voltmeter across the resistor, and then connect the current meter in series with the resistor and switch. Make sure the switch is left open.

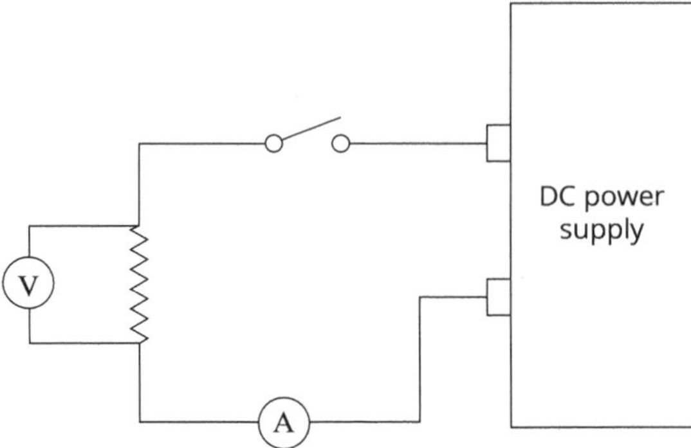

3 Turn on your power supply, and set the output voltage of the power supply to 5 V. Record this value in Table 1 in the Data and Analysis section.

4 Close the switch, and then record the voltage and current reading in Table 1 in the Data and Analysis section. After recording the values, open the switch.

5 Add the second resistor in series with the first, and then move the voltage sensor leads to measure the total voltage across both resistors. Close the switch, and record the voltage and current reading in Table 1 in the Data and Analysis section.

6 Move the voltage meter leads to now measure the voltage across the first resistor by itself, and record the voltage in Table 1 in the Data and Analysis section.

7 Move the voltage meter leads to measure the voltage across the second resistor by itself, and record the voltage in Table 1 in the Data and Analysis section. After recording the values, open the switch.

8 Add the third resistor in series with the first, and then repeat the previous steps recording the voltage and current across all three resistors, and then the voltage for each individual resistor in the circuit. Record the values in Table 1 in the Data and Analysis section.

9 Open the switch before moving on to the next section.

Resistors in parallel

1 With the power supply off, connect the first resistor and the switch (open) in series to the power supply using the patch cords.

2 Connect the voltage meter across the resistor, and then connect the current meter in series with the resistor and switch. Be sure the switch is open.

3 Turn the power supply on, and set the output voltage to 5 V.

4 Close the switch, and then record the voltage and current reading in Table 2 in the Data and analysis section. After recording the values, open the switch.

5 Add the second resistor in parallel with the first as shown below.

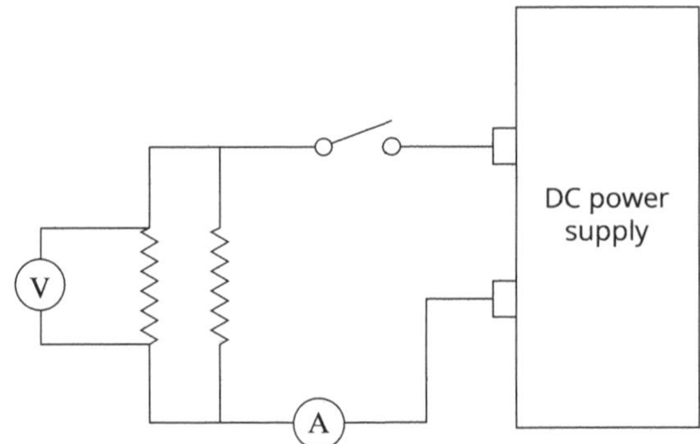

6 Close the switch, and record the voltage and current reading in Table 2 in the Data and analysis section. After recording the values, open the switch.

 ISBN 978 1 4886 1935 9

7 Move the current meter leads to measure the current through the first resistor by itself.

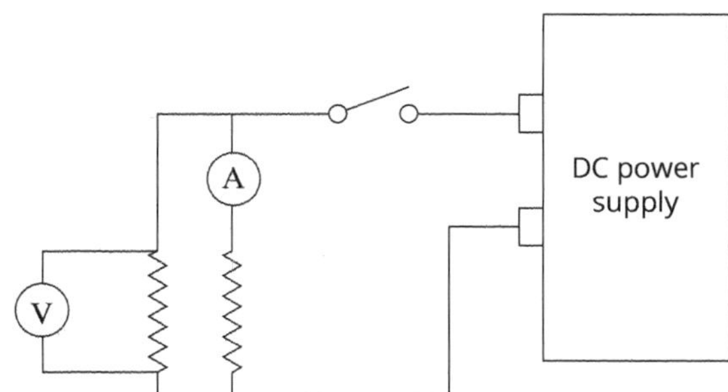

8 Close the switch, and record the current in Table 2 in the Data and analysis section. After recording the value, open the switch.

9 Move the current meter leads to measure the current through the second resistor, close the switch, and record the current reading in Table 2 in the Data and analysis section. After recording the value, open the switch.

10 Add the third resistor in parallel with the first two, and repeat the previous steps recording the voltage and current across all three resistors, then the current for each individual resistor in the circuit. Record the values in Table 2 in the Data and analysis section.

11 Open the switch after you have made the final measurement.

DATA AND ANALYSIS

Calculate and enter the theoretical resistor rating from the values of the resistors used for each circuit in each of Tables 1 and 2.

Calculate the measured value for resistance for each circuit using $R = \frac{V}{I}$ and enter the values into the corresponding columns in Tables 1 and 2.

Voltage from power supply: _____ V

TABLE 1 Series circuit

Number of resistors	Resistor	Voltage (V)	Current (A)	$\frac{\text{Voltage}}{\text{Current}}$	Resistance (Ω)
1	First				
2	First				
	Second				
Total voltage for two resistors:		V			
3	First				
	Second				
	Third				
Total voltage for three resistors:		V			

ISBN 978 1 4886 1935 9

TABLE 2 Parallel circuit

Number of resistors	Resistor	Voltage (V)	Current (A)	Voltage/Current	Resistor Rating (Ω)
1	First				
2	First				
	Second				
Total current for two resistors:		V			
3	First				
	Second				
	Third				
Total current for three resistors:		V			

CONCLUSION

1 For the series circuit, the voltage was initially measured for each component. Why wasn't the current being measured for each component as well?

2 For the parallel circuit, the current was measured for each component. Why wasn't the voltage being measured as well?

3 For each circuit, comment on how the measured resistance compares with the theoretical resistance for each resistor.

RATING MY LEARNING	My understanding improved	Not confident ◄──► Very confident ○ ○ ○ ○ ○	I answered questions without help	Not confident ◄──► Very confident ○ ○ ○ ○ ○	I corrected my errors without help	Not confident ◄──► Very confident ○ ○ ○ ○ ○

 ISBN 978 1 4886 1935 9

PRACTICAL ACTIVITY 4.5

Magnetic field of a coil

Suggested duration: 40 minutes

INTRODUCTION

The use of coils in electromagnetic devices varies from transformers allowing electrical supply over long distances to pickup coils in electric guitars, voice coils in loudspeakers and coils in electric motors. Coils, or solenoids as they are also known, are a key enabling technology of modern society.

The strength of the magnetic field inside a coil depends upon the number of turns of the wire, the magnitude of the current in the wire, and the radius of the coil. i.e.

$$B = N\frac{\mu_0 I}{2\pi r}$$

where:

B is the magnitude of the magnetic field strength (in T)

N is the number of turns

μ_0 is the permeability of free space, or magnetic constant ($\mu_0 = 1.257 \times 10^{-6}\,\mathrm{N\,A^{-2}}$)

I is the current (in A)

r is the radius of the coil (in m).

The magnetic field at a point well inside a solenoid is uniform and independent of the length or diameter of the solenoid. Instead, it is dependent upon the number of turns of the coil (N) per unit length (L) of the solenoid and the equation becomes:

$$B = \frac{\mu_0 N I}{L}$$

The more tightly wound the wire around a solenoid, the greater the magnetic field. Solenoids can be used to generate nearly uniform magnetic fields, similar to a bar magnet.

MATERIALS

- data collection system
- current sensor
- magnetic field sensor
- 3 coils of varying turns but the same radius
- metre ruler
- DC power supply (10 V 1 A minimum)
- 3 patch cords with banana plugs

PURPOSE

To experimentally investigate (a) the relationship between the magnetic field strength inside a coil and the magnitude of the current carried through the coil wire, and (b) the relationship between the number of turns of wire in a coil and the magnetic field strength within the coil.

This investigation is best performed with a data collection system. However, if this is unavailable, reasonable comparative results can be obtained using an ammeter and a magnet suspended on a string. Record the deflection of the magnet as an alternative to magnetic field strength.

Strong magnetic fields will be developed within and around the coil. Keep the coil away from any sensitive electronic equipment such as phones, tablets and computers.

Keep liquids away from the power supply.

PROCEDURE

1 Start a new experiment in the data collection system and connect both magnetic field sensor and current sensor. (Refer to the note in the Materials section if these are not available to you).

2 Set up displays in the data collection software to show digital readings to monitor current and magnetic fields.

Part 1: Varying the number of turns in the coil

1 Measure the radius of the coils and record this value in the Data and analysis section.

2 Using the patch cords, connect the negative terminal from the power supply directly to one of the terminals on the first coil. Connect the positive terminal from the power supply to the positive terminal on the current sensor, and then connect the negative terminal on the current sensor to the second terminal on the coil to complete the circuit.

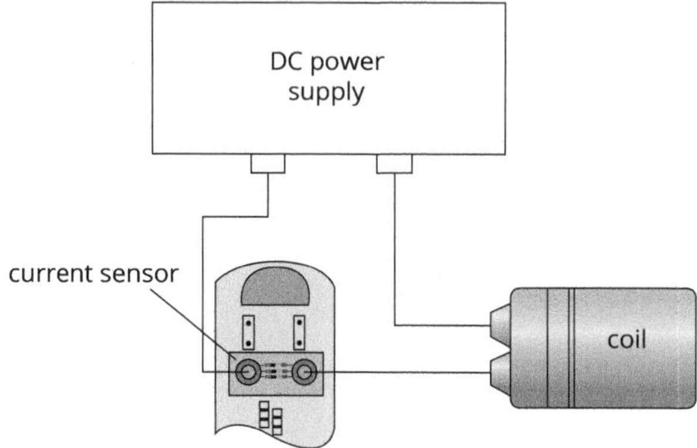

3 Position the tip of the magnetic field sensor inside the coil at the centre so that the probe of the sensor is perpendicular to the plane of the coil.

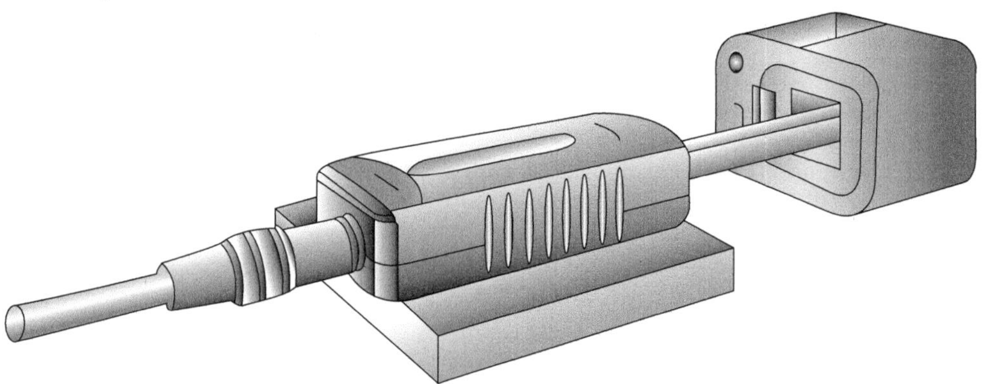

4 Turn on the power supply, and adjust the voltage until the current measured by the current sensor is approximately 0.5 A.

5 Record the current value in the Data and analysis section.

6 Record the number of turns in the coil and magnetic field strength in Table 1 in the Data and analysis section.

7 Turn off the power supply, disconnect the coil, and replace it with the next coil.

8 Reposition the coil and magnetic field sensor, and then turn the power supply back on. Adjust the voltage until the current is the same magnitude as was used with the previous coil.

9 Record the number of turns in the coil and the magnetic field strength in Table 1 in the Data and analysis section.

10 Turn off the power, disconnect the coil, and replace it with the last coil.

11 Reposition the coil and magnetic field sensor, and then turn the power supply back on. Adjust the voltage until the current is the same magnitude as was used with the previous coils.

12 Record the number of turns in the coil and the magnetic field strength in Table 1 in the Data and analysis section.

Part 2: Varying the current through the coil

1 Using the same set-up as in Part 1 and using the first coil, turn the voltage down to zero on the power supply and turn it on. Record the magnetic field strength in Table 2 in the Data and analysis section.

2 Increase the voltage slightly until the current increases about 0.1 A. Record the magnetic field strength and current values in Table 2 in the Data and analysis section.

3 Adjust the voltage until the current increases approximately another 0.1 A (i.e. approximately 0.2 A in total). Once again, record the magnetic field strength and current values in Table 2 in the Data and analysis section.

 ISBN 978 1 4886 1935 9

4 Continue collecting data points every 0.1 A through to approximately 1.0 A, and then stop data collection and turn the power supply off.

 Note the limit of the current sensor or ammeter carefully. Do not exceed this rating with the investigation of current range.

DATA AND ANALYSIS

Calculate the expected magnetic field for each coil using the theoretical formula and complete Table 1.

Compare the measured values to the calculated values for magnetic field strength within each coil. Calculate the percentage difference and add it to Table 1.

Sketch a graph of magnetic field strength versus number of turns on the grid below.

Part 1: Varying the number of turns in the coil

Current in the coils: ____________ A

Radius of the coils: ____________ m

TABLE 1

Number of turns (N)	Measured magnetic field B (T)	Calculated magnetic field B (T)	% difference

Calculate the expected magnetic field for each coil using the theoretical formula and complete Table 1.

Compare the measured values to the calculated values for magnetic field strength within each coil. Calculate the percentage difference and add it to Table 1.

Sketch a graph of magnetic field strength versus current on the grid provided.

ISBN 978 1 4886 1935 9

Part 2: Varying the current through the coil

Radius of the coils: ________ m

Number of turns: ________

TABLE 2

Current (A)	Measured magnetic field B (T)	Calculated magnetic field B (T)	% difference

 ISBN 978 1 4886 1935 9

CONCLUSION

1 What variables can be changed in order to investigate the effect on magnetic field strength in a coil?

2 Describe the relationship between magnetic field strength and the number of turns and current in the coil with reference to your graphs.

3 How can any differences between measured and calculated values for the magnetic field strength be accounted for? Explain and include suggestions for improving the measured results.

RATING MY LEARNING	My understanding improved	Not confident ◄──► Very confident ○ ○ ○ ○ ○	I answered questions without help	Not confident ◄──► Very confident ○ ○ ○ ○ ○	I corrected my errors without help	Not confident ◄──► Very confident ○ ○ ○ ○ ○

DEPTH STUDY 4.1

Magnetic fields—the invisible enablers

Suggested duration: 3.5–4 hours, including data collection, analysis and report

INTRODUCTION

Usually invisible to the human eye except when highlighted by natural phenomena such as solar flares and the aurora australis, magnetic fields are as ubiquitous on the Sun and most of the planets in our solar system as gravitational fields. The Earth's magnetic field provides a shield from solar radiation that has allowed life to evolve on Earth. Many animals rely on magnetic fields for navigation, and the interaction between electrical and magnetic fields is a fundamental technology of modern society.

This depth study requires you to question the role of magnetic fields in a natural phenomenon or modern technology. You will develop an inquiry question that requires research to develop an informed hypothesis, plan a research investigation and analyse second-hand data and information from appropriate sources and use problem-solving techniques to determine the validity of the sources. You will evaluate processes, claims and conclusions by considering the quality of the available evidence, and use reasoning to construct scientific arguments. You will process the data and information in order to communicate your findings in a short individual presentation that includes your initial research, justification of your sources and development of knowledge and understanding of the topic. Your presentation will use appropriate scientific notations and nomenclature, and appropriately apply and use scientific language that is suitable for the audience and context.

Artist's impression of the aurora borealis, or northern lights, by Frederick C. Bakewell (1853). An aurora occurs when charged particles from the Sun are drawn along the Earth's magnetic field to the polar regions. Light is emitted when these particles collide with molecules and atoms in the upper atmosphere.

In this activity you will use secondary resources to report on a natural phenomenon or modern application of magnetic fields, including emerging technologies, clearly distinguishing the specific role of magnetic fields and the relevant physics. Your report will take the form of an individual documentary, media report or other short visual presentation.

PURPOSE

To research a specific topic relating to a natural phenomenon or modern application of magnetic fields using second-hand data, develop an appropriate hypothesis that explains the role of magnetic fields in your topic, and report on your knowledge and understanding of the topic based on an analysis of the second-hand data.

 ISBN 978 1 4886 1935 9

TOPIC REQUIREMENTS AND CONSTRAINTS

Your research must be conducted individually.

The topic chosen must have a natural or technological application of magnetic fields as the primary area of research.

The topic must allow for the development of and answer to one clear inquiry question.

A topic that has a broad range of resources available is likely to prove better than one where the resources are limited.

QUESTIONING AND PREDICTING

1 Will your topic look at a natural phenomenon or a recent or developing technological application of magnetic fields?

2 Phrase your topic as a question. For example, 'How does the Earth's magnetic field protect the Earth from solar radiation?' You may want to consult a few resources first.

3 How do magnetic fields apply to your question? List 2 or 3 possible questions that would help answer your main question. For example, 'Why is solar radiation stopped by the Earth's magnetic field?', 'How does solar radiation interact with magnetic fields?', 'Does this apply to all solar radiation?'

4 Do some initial research around answering your initial questions. Rephrase your questions to make them as clear and specific as possible. For example, 'How do charged particles interact with magnetic fields?', 'What charged particles can be found in solar radiation?', 'What effect does the Earth's magnetic field have on charged particles in solar radiation?'

5 Look back at your initial topic question. Construct a hypothesis that can be applied to answer your question. This becomes your working hypothesis and should summarise the answer to your main research question. It will most likely change after some further research.

ANALYSING DATA AND INFORMATION

6 Look back at your original question. Does it need rephrasing in light of your research? Restate your question as you will present it.

7 Working in a small group, evaluate other students' research topics. What are the strengths of their research questions? How could you improve your own research question?

PROCESSING DATA AND INFORMATION

8 Carry out the remainder of your research, summarising key points as you proceed so that your topic has a clear focus and structure. Use a tree diagram like the one in the figure below to link your research questions back to your original hypothesis.

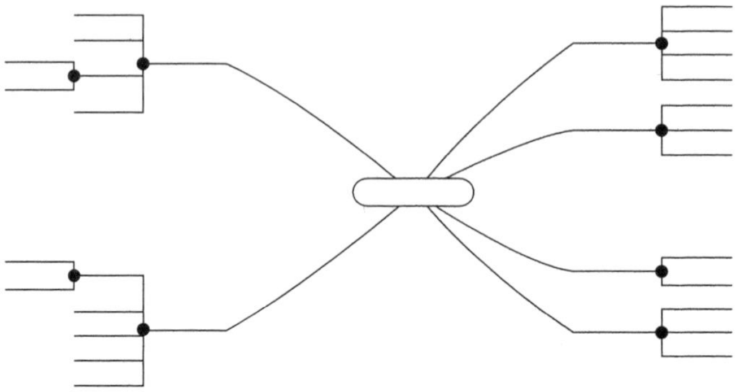

 ISBN 978 1 4886 1935 9

COMMUNICATING

Communicate your findings in the form of an individual documentary, media report or other short visual presentation. Evaluate your research using the following questions.

9 Were you successful in answering your original research question? Did you need to rephrase it?

10 Do your research sources all agree? Are there any dissenting claims or explanations?

11 To what extent were you able to verify or reference any dissenting claims? How are these presented in your research?

Multiple choice

1 Which one or more of the following terms are equivalent to the energy gained or lost by charge in a circuit?

A EMF

B amperage

C voltage

D resistivity

2 The force between two equal charges, q, is F newtons when they are d cm apart.

How far apart will charges of q and $2q$ need to be so that the force is the same?

A $\frac{d}{4}$

B $\frac{d}{2}$

C $\sqrt{2}d$

D $2d$

3 Two aluminium spheres have charges of $-q$ and $+2q$ and experience a force of F newtons when they are d cm apart. If they are momentarily touched together and then moved $0.5d$ cm apart, how large will the force between them become?

A F

B $\frac{F}{2}$

C $\frac{F}{8}$

D $\frac{3F}{4}$

4 Given that the red end of a compass needle points north, which two of the following answers are true?

A The red end is a magnetic south pole.

B The red end is a magnetic north pole.

C The Earth's north pole is a magnetic south pole.

D The Earth's north pole is a magnetic north pole.

5 How much current is drawn by a 6.0 W globe connected to two 1.5 V batteries placed in series?

A 0.5 A

B 4.0 A

C 9.0 A

D 2.0 A

6 The magnetic field strength at a point d away from a long straight wire carrying a current I is B. How strong will the field be $2d$ away from the wire when it carries a current of $2I$?

A B

B $\frac{B}{2}$

C $2B$

D $4B$

Short answer

7 A speck of dust with a charge of $+2.5 \times 10^{-10}$ C is suspended midway between a pair of large horizontal metal plates 1.0 cm apart which have a potential difference of 400 V between them.

a Which plate is positively charged?

b Calculate the mass of the dust speck if the gravitational acceleration g is $10\,\text{m}\,\text{s}^{-2}$.

c What net force will the dust speck experience if the two plates are moved twice as far apart?

d What will happen then?

8 Mary works out that the 12 V battery in her car is connected via a single switch to two heater elements (each with a resistance of 3.0 Ω) in parallel, which are used to heat the driver and passenger seats.

a Draw a labelled diagram to represent this circuit.

 ISBN 978 1 4886 1935 9

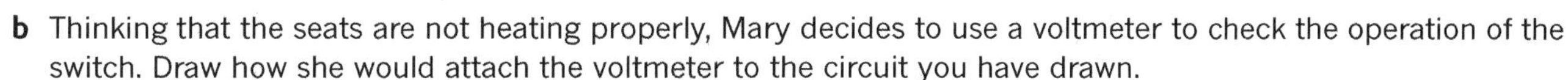

b Thinking that the seats are not heating properly, Mary decides to use a voltmeter to check the operation of the switch. Draw how she would attach the voltmeter to the circuit you have drawn.

c When the switch is open, what value should the voltmeter read?

d When the switch is closed, what value should the voltmeter read?

e Mary wants an extra switch so that each seat can be warmed individually. Draw the new circuit, showing where that switch should be placed.

f How much power would normally be used to heat one seat?

9 Consider the circuit below. The current through the 9.0 Ω resistor is 400 mA.

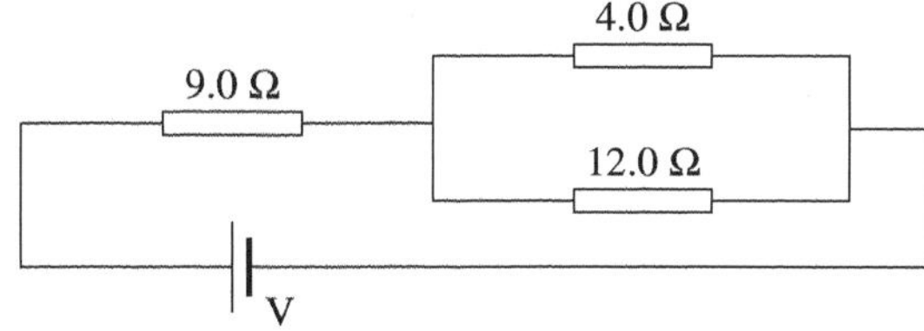

a What is the power produced in the 9.0 Ω resistor?

b What is the total resistance of the circuit?

c Calculate the voltage V of the battery.

d What power is produced in the $4.0\,\Omega$ resistor?

10 Two charges P (3.0×10^{-8} C) and Q (6.0×10^{-8} C) are placed 40 cm apart as shown in the diagram below.

a Calculate the force acting on Q.

b What is the size of the electric field at the point midway between P and Q?

c What is the size and direction of the electric field at point R?

 ISBN 978 1 4886 1935 9

Extended response

11 Two electrical devices have current–voltage characteristics as shown in the diagram below.

Device P

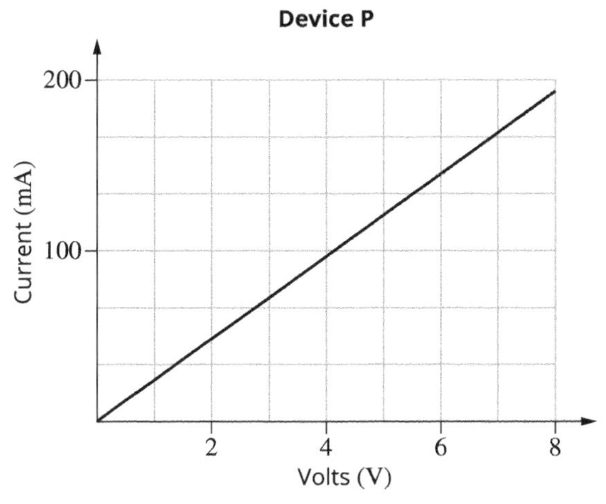

Device Q

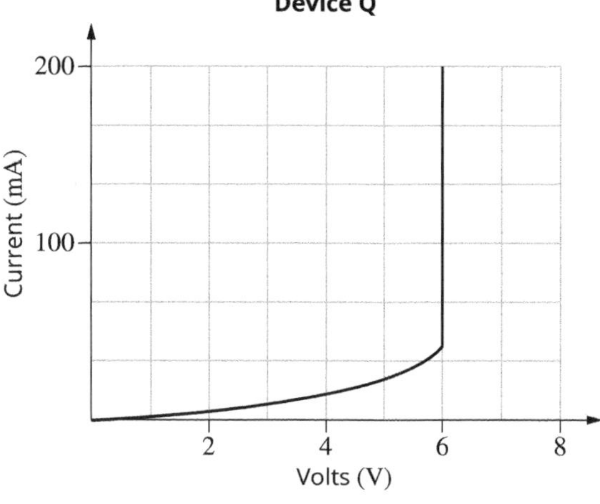

a Which of the two devices, P or Q, obeys Ohm's law?

b Determine the resistance of that device.

c What is the resistance of the other device when a current of 200 mA flows through it?

d The two components are now connected in a series circuit as shown. The current at point K is 100 mA. What is the current at point L?

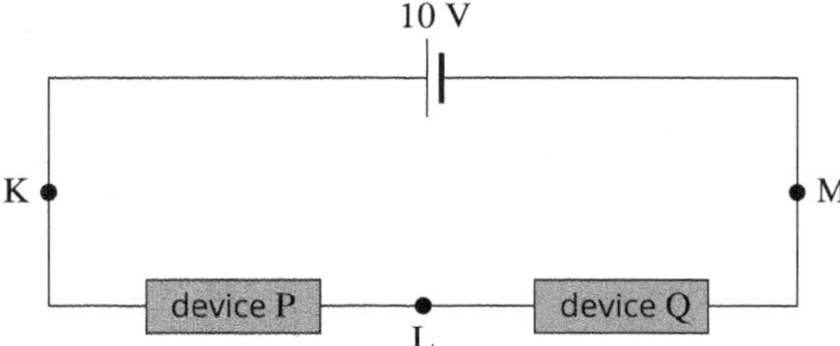

e Calculate the potential difference between points K and L.

f Calculate the potential difference between points L and M.

g Calculate the current through the battery if an extra device P is connected in series with the first device P.

12 Jess sets up the following circuit in her spare time, using a 6.0 V battery and three identical light globes. She also has an ammeter and a voltmeter.

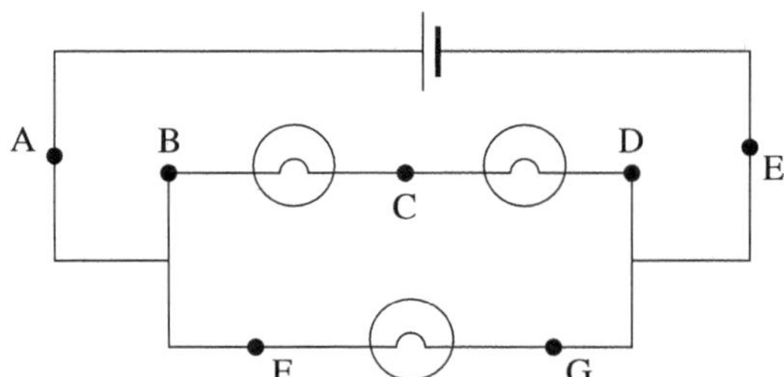

a Describe exactly what she would have to do to measure the potential difference across the lower globe.

b Jess recorded a current of 240 mA through the lower globe. Describe exactly how she did this.

c Fill in the missing values that Jess should find when she makes the following measurements.

Potential difference between A and E	
Potential difference between C and D	
Current through point C	160 mA
Potential difference between D and E	
Current through point A	

d Jess thought that the current through point C would be 120 mA. Why would she have expected that value?

e Explain why this value is actually 160 mA.

 ISBN 978 1 4886 1935 9

Notes

Notes

ISBN 978 1 4886 1935 9

Notes

 ISBN 978 1 4886 1935 9